T0340324

Project Management Essentials

Analytics and Control Series

Series Editor:
Adedeji B. Badiru, Air Force Institute of Technology,
Dayton, Ohio, USA

Decisions in business, industry, government, and the military are predicated on performing data analytics to generate effective and relevant decisions, which will inform appropriate control actions. The purpose of the focus series is to generate a collection of short-form books focused on analytic tools and techniques for decision-making and related control actions.

Mechanics of Project Management
Nuts and Bolts of Project Execution
Adedeji B. Badiru, S. Abidemi Badiru, and I. Adetokunboh Badiru

The Story of Industrial Engineering
The Rise from Shop-Floor Management to Modern Digital Engineering
Adedeji B. Badiru

Innovation
A Systems Approach
Adedeji B. Badiru

Project Management Essentials
Analytics for Control
Adedeji B. Badiru

For more information on this series, please visit: https://www.routledge.com/Analytics-and-Control/book-series/CRCAC

Project Management Essentials
Analytics for Control

Adedeji B. Badiru

CRC Press
Taylor & Francis Group
Boca Raton London New York

CRC Press is an imprint of the
Taylor & Francis Group, an **informa** business

First edition published 2021
by CRC Press
6000 Broken Sound Parkway NW, Suite 300, Boca Raton, FL 33487-2742

and by CRC Press
2 Park Square, Milton Park, Abingdon, Oxon, OX14 4RN

Library of Congress Cataloging-in-Publication Data
Names: Badiru, Adedeji Bodunde, 1952- author.
Title: Project management essentials : analytics for control / Adedeji B. Badiru.
Description: First edition. | Boca Raton : CRC Press, 2021. |
Series: Analytics and control | Includes bibliographical references and index.
Identifiers: LCCN 2020041949 (print) | LCCN 2020041950 (ebook) |
ISBN 9780367431181 (hbk) | ISBN 9781003004172 (ebk)
Subjects: LCSH: Project management.
Classification: LCC HD69.P75 B3276 2021 (print) | LCC HD69.P75 (ebook) |
DDC 658.4/04—dc23
LC record available at https://lccn.loc.gov/2020041949
LC ebook record available at https://lccn.loc.gov/2020041950

ISBN: 978-0-367-43118-1 (hbk)
ISBN: 978-0-367-71059-0 (pbk)
ISBN: 978-1-003-00417-2 (ebk)

Typeset in Times
by codeMantra

Dedicated to the memory of Dr. Gary E. Whitehouse, whose guiding hands pointed me in the direction of project optimization and control.

Contents

Preface

Project management represents an excellent basis for achieving goals. Project management is around us, whether we realize it or not. With the author's hypothesis that everything is a project, it means that project management is needed every time and everywhere. The purpose of this book is to present a focused approach to project management for all readers. The particular focus is on simple analytics for project control, using the systems framework of the DEJI® systems model of Design, Evaluation, Justification, and Integration®. The overriding theme of this book is that every pursuit can be organized as a project. The phrase "project management is everything" implies that project management permeates every pursuit. Similarly, the phrase "everything is project management" implies that every success is dependent on project management.

The integrated approach of this book covers the concepts, tools, and techniques of project management, from both qualitative and quantitative perspectives, but with a systematic structure of project design, project evaluation, project justification, and project integration. The elements of the Project Management Body of Knowledge provide a unifying platform for the topics covered in this book. This book is intended to serve as a reference book for everyone, since everyone is engaged in project management, whether formal or informal. It can serve as a useful reference for project analysts, project owners, project researchers, project management teachers and students, project entrepreneurs, and project policy-makers. The goal of this focus book is to provide a concise framework for systems-based project analytics and control. For a more comprehensive treatment, readers are referred to the author's other project management books.

Acknowledgments

Once again, I thank Cindy Carelli and Erin Harris for their teamwork, commitment, and gentle nudging for the completion of this Focus book. I also thank my family for putting up with my frequent disappearance into writing solitude. I particularly appreciate my granddaughter's (Alexa's) acclamation that grandpa is writing many words. Yes, indeed, many words are required to complete a book manuscript.

Author

Adedeji B. Badiru is a professor of Systems Engineering at the Air Force Institute of Technology (AFIT), Dayton, Ohio. He is a registered professional engineer and a fellow of the Institute of Industrial Engineers as well as a fellow of the Nigerian Academy of Engineering. He has a BS degree in Industrial Engineering, MS in Mathematics, and MS in Industrial Engineering from Tennessee Technological University, Cookeville, TN, USA, and a PhD in Industrial Engineering from the University of Central Florida, Orlando, FL, USA. He is the author of several books and technical journal articles.

Project Management with a Systems Framework

> *If it is a hard nut to crack, please, by all*
> *means, use a systems sledge hammer.*
> – Adedeji Badiru

SYSTEMS VIEW OF PROJECT MANAGEMENT

The bottom line of effective project management can be summarized as follows: The longer a project lingers in execution, the more it will be subject to inefficiency, wasteful acts, inconsistency, abuse, fraud, or corruption. For capital projects, a good strategy is to get it started and get it done expeditiously. A systems approach facilitates effective project management.

It is a systems world these days, and wealth and survival of nations are built on the power of executing projects successfully. Projects, both in the public and private sectors of the economy, can benefit from the project management methodology advocated in this book.

There are hundreds of project management books. So, what could be new in writing another project management book? What is new is not necessarily "new and unknown topics." What is new is how what is already known could be structured into a systems-based framework to guide a project onto the path of assured success. That is the premise of this book. The easy temptation is to believe that what is needed to be said about project management has already been said before and elsewhere. But the fact is that what has been said previously may need to be harnessed into a systematic structure for a path to success.

Projects take place within organizational structures. Even home-based personal projects are within the context of home organizational structures, representing a system of interconnected elements. Some considerations in this regard include project commitment, project leadership, performance measures, team cohesion, mutual trust, work satisfaction, level of personal engagement, and work processes.

Little drops of project trouble can make a mighty pile to derail a project. Project scenarios don't have to be catastrophic before they become insurmountable. Sometimes, the pile up of many little issues, each one alone being minor and harmless, can add up to become beyond repair. That is the power of coalescing little elements of a system, whose impact can be positive or negative. The methodology of a systems approach is to channel that system's power positively to operate in favor of the project.

By logical reasoning, everything is a system. There is a system in everything. Everything is a project. There is a project in everything. Consequently, a successful project is that which is executed from a systems perspective. Project management represents an excellent basis for achieving goals. Project management is around us, whether we realize it or not. With the author's hypothesis that everything is a project, it means that project management is needed every time and everywhere. The purpose of this book is to present a focused approach to project management for all readers.

IMPACT OF COVID-19 ON PROJECT MANAGEMENT

It is important to recognize and take heed of the impacts, real, perceived, or anticipated, that the 2020 emergence of the COVID-19 pandemic is already having or will have on any project environment and the practices of project management. Nothing is going to remain the same in what we previously knew of the project environment. Some practices will evolve into something else. Other practices will varnish forever. Due to the increased fast global connectivity, we are now more prone to a fast spread of any pandemic, but we are less prepared to take remedial actions.

In as much as humans still manage projects, new cases of COVID-19 may upend the long-held expectations of conventional project management with respect to workforce development, workforce management, and workforce preservation. For this reason, a systems approach is needed more now than ever before. A systems approach facilitates more flexible, adaptive, and responsive reactions to new developments in the project environment. The efficacy of this book lies on the necessity of developing new systems-based views and approaches to project management. The workforce of the future will be more and more online, working remotely. COVID-19 gives us an opportunity for a kick-start to the future work environment. A case example of a community commentary on workforce development during and post-COVID-19 (Badiru and Barlow, 2020) is presented at the end of this chapter.

INTRODUCTION TO THE DEJI SYSTEMS MODEL®

The particular focus of this book is on using simple and familiar analytics (not necessarily mathematical or statistical) for project control, based on the systems framework of the DEJI Systems Model® of **Design**, **Evaluation**, **Justification**, and **Integration**. With this systematic framework, the theme of analytics for control, embodied in this book's title, can be accomplished.

Analytics in the context of the application of the DEJI Systems Model does not need to be overly complicated or solely quantitative. In some cases, qualitative analytics can be as effective as quantitative analytics. Good examples include a simple counting of how many stakeholders have buy-in into the project and to what extent, an analysis of authorization and empowerment available for the project, the resource base available for the project, a documentation of the strength of compliance with regulatory guidelines for the project, a review of the growth opportunities available for project participants, and a listing of the hierarchical points of decision that the project is subjected to. The analytics of "eclipsity," which this book defines as a measure of overshadowing (eclipsing) project potentials with incompetencies, can be identified and mitigated by a broad systems approach. Analytics doesn't have to be esoteric. Simple charts portraying trends can help present facts and defuse fears.

Control in the context of the application of the DEJI Systems Model does not need to be draconian. Simple actions such as making preemptive decisions can be effective in ensuring that a project stays on track. Where and when corrective actions are needed, they are implemented as routine measures of project progress rather than brusque shifts in project directions. Mentoring and managing human resources proactively can serve as preemptive control measures in a project. With a systems approach, the project team can use innovative and creative strategies upfront to ensure that the project execution is aligned with its expected outcomes. As the saying goes, "a picture is worth a thousand words," so it is that data analytics charts can be worth score of debate about project control.

The overriding theme of the book is that every pursuit can be organized as a project and every project can be managed inside-out, based on its inherent subsystems. The phrase "project management is everything" implies that project management permeates every pursuit. Similarly, the phrase "everything is project management" implies that every success is dependent on project management. As a personal case example, colleagues and observers often ask me how I could be, seemingly effortless, involved in so many things, often in parallel, and yet seem to be successful with most and loving it. My simple answer is always "project management." I am convinced it works as a personal approach. The key is to prioritize and pursue only what matters. Don't take on more than can be done. Don't promise what cannot be delivered. Eliminate the fluff and focus on the grit. Even in the corporate world, over-diversification and over-extending available resources often lead to corporate business failures. The integrated approach advocated here covers the concepts, tools, and techniques of project management, from both qualitative and quantitative perspectives, but with a systematic structure of project design, project evaluation, project justification, and project integration. The elements of the Project Management Body of Knowledge provide a unifying platform for the topics covered in this book. This book is intended to serve as a reference book for everyone, since everyone is engaged in project management, whether formal or informal. This book can serve as a useful reference for project analysts, project owners, project researchers, project management teachers and students, project entrepreneurs, and project policy-makers. The thematic layout of this book is based on the following structure:

- Systems View of Project Management
- Project Design
- Project Evaluation
- Project Justification
- Project Integration.

This chapter addresses a general introduction to the application of systems thinking to project management. The four chapters that follow sequentially focus on project design, project evaluation, project justification, and project integration, which form the elements of the DEJI® systems model. Thus, everything that we may already know and embrace in project management can be accommodated in one of more stages of the DEJI® Systems Model, which is illustrated in Figure 1.1.

The various elements contained in DEJI Systems Model describe most of the components typically found in any project. Looking at the model from inside out, the following components are described:

Practicality: A project must be assessed for its practicality. Not all projects that are conceived are practical in implementation.

Sustainability: In our modern era of concerns with climate change, every project should be sensitive to the issues of environmental sustainability. The resources needed and consumed by a project are the same resources of concern and contention in sustainability movements. Apart from environmental sustainability, literal sustainability of a project relates to the project's ability to survival, continue, and be sustainable to the end of its life cycle.

Desirability: Not all projects that are possible, practical, and sustainable are desirable. From a systems perspective, projects should be evaluated on the basis of how desirable they are to the stakeholders and the observers.

Agility: For a project, a measure of agility is the project's ability to adapt and leverage new developments and opportunities in its environment. Agile project management is a new body of pursuit that has evolved in

FIGURE 1.1   DEJI® Systems Model for project management essentials.

project management communities in recent years. Agility can be defined and approached in different ways by different organizations. However, the common elements often revolve around collaborative efforts of self-organizing and cross-functional teams with a focus on customer requirements.

Feasibility: The multiple dimensions of a project should be assessed with respect to the pertinent angles of feasibility. A project that may be feasible in technical terms may not be feasible in economic terms.

Affordability: Some projects that may be grandiose in concept may not be affordable in implementation. If a project is not affordable, it will not be feasible for long.

Metrics: Planning for a project may be very good, but its control may be lacking due to the absence of metrics. Metrics provide the rubrics for assessing how well a project is meeting its goals and objectives. Metrics for project management should be identified and agreed upon at the outset. Otherwise, discord and disconnect may develop down the line.

All these internal elements are embodied within the overall framework of design, evaluation, justification, and integration. On the outer circle of the model, we have the elements consisting of defining the end goal, gathering evidence, articulating conclusions, assessing utility, engaging stakeholders, and focusing on project implementation. Each organization and each project team may have additional or diverse elements to include in the concentric circles of the DEJI Systems Model. The benefit of the model is that it provides a systematic framework for coalescing options and opportunities within the project environment.

CONGRUENCE WITH BLOCKCHAIN

A digital implementation makes things more traceable and recordable, thereby minimizing inefficiency, ineptitude, and ineffectiveness. The popularity of the methodology of blockchain, as a digital tool, can be leveraged for the advancement of the application of DEJI Systems Model to general project management. Blockchain is defined simply as an open, decentralized, and distributed ledger technology that records the origin, attribution, derivation, source, or background of an organization's digital asset. It, essentially, implies taking a systems view of how digital assets are linked, utilized, and promoted throughout an organization to the benefit of managing assets and projects better. Blockchain is a digitally intensive technology, which many information technology companies are embracing and developing. Blockchain originated in the cryptocurrency domain as a way to track transactions in a financial bitcoin domain to prevent double spending by providing a payment trail. The complexity of the technology behind blockchain makes it rare and elusive in application. In essence, it is a technically demanding creation and manipulation of a higher-level database. Thus, a simpler adaptation for general business application is desired. Because of the linkages in the database, the system has an exponentially growing list of records, which are called blocks, hence the formation of a chain of blocks. So, in the parlance of a systems structure, a blockchain can be viewed as a collection of interrelated data records, whose overall value is determined by the output of the blockchain rather than the individual contents of records. This makes a blockchain amenable for design, evaluation, justification, and integration for project execution purposes. By its design, a blockchain is impervious to random modification of the data. This is a good control measure. It is typically managed by peer-to-peer networking protocol for inter-node communication and validation of new data blocks. The concept is useful for framing project execution within a systems structure.

GENERAL SYSTEMS FRAMEWORK

In general, a system is defined as a collection of interrelated elements, whose collective output, together in unison, is higher than the sum of the individual outputs. A system is represented as consisting of multiple parts, all working together for a common purpose or goal. Systems can be small or large, simple or complex. Small devices can also be considered systems. Systems have inputs, processes, and outputs. Systems are usually explained using a model for a visual clarification inputs, process, and outputs. Systems engineering is the application of engineering tools and techniques to the solutions of multifaceted problems through a systematic collection and integration of parts of the problem with respect to the life cycle of the problem. It is the branch of engineering concerned with the development, implementation, and use of large or complex systems. It focuses on specific goals of a system considering the specifications, prevailing constraints, expected services, possible behaviors, and structure of the system. It also involves a consideration of the activities required to ensure that the system's performance matches specified goals. Systems engineering addresses the integration of tools, people, and processes required to achieve a cost-effective and timely operation of the system. Some of the features of this book include considerations of multi-faceted problems, holistic views of problem domains, applications to both small and large problems, decomposition of complex projects into smaller manageable chunks, direct considerations for the pertinent constraints that exist within a project, systematic linking of inputs to goals and outputs, and explicit treatment of the integration of tools, people, and processes.

The end result of using a systems approach is to integrate a solution into the normal organizational process. For that reason, the DEJI Systems Model is desired for its structured framework of Design, Evaluation, Justification, and Integration. The International Council on Systems Engineering (INCOSE) developed the INCOSE Systems Engineering Competency Framework (INCOSE SECF), which represents a world view of five competency groupings with 36 competencies central to the profession of Systems Engineering. This includes evidence-based indicators of knowledge, skills, abilities, and behaviors across five levels of proficiency. INCOSE SECF supports a wide variety of usage scenarios including individual and organizational capability assessments. It enables organizations to tailor and derive their own competency models that address their unique challenges. The DEJI Systems Model embraces the INCOSE SECF framework and complements many of the elements contained in the framework. The five competency groupings and their respective competencies are summarized below:

- Systems Competency
  1. Information Management
  2. Planning
  3. Monitoring and Control
  4. Risk and Opportunity Management
  5. Business and Enterprise Integration
  6. Decision Management
  7. Concurrent Engineering
  8. Configuration Management
  9. Acquisition and Supply
- Professional Competency
  1. Communications
  2. Facilitation
  3. Ethics and Professionalism
  4. Coaching and Mentoring
  5. Technical Leadership
  6. Emotional Intelligence
  7. Negotiation
  8. Team Dynamics

- Core Systems Engineering Principles Competency
  1. General Engineering
  2. Systems Thinking
  3. Capability Engineering
  4. Critical Thinking
  5. Systems Modeling and Analysis
  6. Life Cycles
- Integrating Competency
  1. Quality
  2. Finance
  3. Project Management
  4. Logistics
- Technical Competency
  1. Requirements Definition
  2. Systems Architecting
  3. Transition
  4. Operations and Support
  5. Design (for specific needs)
  6. Integration
  7. Interfaces
  8. Verification
  9. Validation.

Each competency in the SECF has five proficiency levels summarized below.

AWARENESS PROFICIENCY

The person displays knowledge of key ideas associated with the competency area and understands key issues and their implications.

SUPERVISED PRACTITIONER PROFICIENCY

The person displays an understanding of the competency area and has some limited experience.

PRACTITIONER PROFICIENCY

The person displays both knowledge and practical experience of the competency area and can function without supervision on a day-to-day basis.

LEAD PRACTITIONER PROFICIENCY

The person displays extensive and substantial practical knowledge and experience of the competency area, and provides guidance to others including practitioners encountering unusual situations.

EXPERT PROFICIENCY

In addition to extensive and substantial practical experience and applied knowledge of the competency area, this individual contributes to and is recognized beyond the organizational or business boundary.

It should be noted that in many professional environments, it is believed that it takes about 15 years of practical experience to become an expert in any particular professional pursuit.

SYSTEMS ATTRIBUTES, FACTORS, AND INDICATORS

In any systems approach to project management, the project team must be cognizant of the attributes, factors, and indicators that fully describe the overall system. The team must look for characteristics of the project that can serve as attributes, factors, and indicators of the desired competencies and proficiencies. Depending on how the project team or the parent organization defines the project environment, some of the elements that can be categorized as attributes, factors, or indicators include teamwork, leadership quality, clarity of direction, clarity of objectives, alignment of goals, resource availability, awareness of relevant industry, caring for people, integrity, consistency, stability of operations, business climate, workforce capabilities, regulatory impositions, industry agreement, customer engagement, public view, social responsibility, financial resources, external business climate, health care infrastructure, investment opportunities, responsiveness, adaptability, operational resiliency, risk management philosophy, portfolio of assets, and so on.

The key is for each project to take advantage of the system around it. What is seen as an attribute in one project scenario may be an indicator for another project scenario within the same project life cycle. Life cycle models vary according to the nature, purpose, use, and prevailing circumstances of a project. Despite an infinite variety in system life cycle models, there is an essential set of characteristic life cycle phases that exists for use in the systems engineering domain. For example, the *Conceptualize* phase focuses on identifying stakeholder needs, exploring different solution concepts, and proposing candidate solutions. The *Development* phase involves refining the system requirements, creating a solution description, and building a system. The *Operational Test & Evaluation* phase involves verifying/validating the system and performing the appropriate inspections before it is delivered to the user. The *Transition to Operation* phase involves the transition to utilization of the system to satisfy the users' needs via training or handoffs. These four life cycle phases are within the scope of any robust systems model.

DEFINITIONS FOR PROJECT SYSTEMS

One definition of systems project management offered by Badiru (2019) is as follows:

> Systems project management is the process of using systems approach to manage, allocate, and time resources to achieve systems-wide goals in an efficient and expeditious manner.

The above definition calls for a systematic integration of technology, human resources, and work process design to achieve goals and objectives. There should be a balance in the synergistic integration of humans and technology. There should not be an overreliance on technology nor should there be an over-dependence on human processes. Similarly, there should not be too much emphasis on analytical models to the detriment of common-sense human-based decisions.

Systems engineering is growing in appeal as an avenue to achieve organizational goals and improve operational effectiveness and efficiency. Researchers and practitioners in business, industry, and government are all clamoring collaboratively for systems engineering implementations. So, what is systems engineering? Several definitions exist. Below is one quite comprehensive definition:

> Systems engineering is the application of engineering to solutions of a multi-faceted problem through a systematic collection and integration of parts of the problem with respect to the lifecycle of the problem. It is the branch of engineering concerned with the development, implementation, and use of large or complex systems. It focuses on specific goals of a system considering the specifications, prevailing constraints, expected services, possible behaviors, and structure of the system. It also involves a consideration of the activities required to assure that the system's performance matches the stated goals. Systems engineering addresses the integration of tools, people, and processes required to achieve a cost-effective and timely operation of the system.

Logistics can be defined as the planning and implementation of a complex task, the planning and control of the flow of goods and materials through an organization or manufacturing process, or the planning and organization of the movement of personnel, equipment, and supplies. Complex projects represent a hierarchical system of operations. Thus, we can view a project system as a collection of interrelated projects all serving a common end goal. Consequently, we present the following universal definition:

> Project systems logistics is the planning, implementation, movement, scheduling, and control of people, equipment, goods, materials, and supplies across the interfacing boundaries of several related projects.

Conventional project management must be modified and expanded to leverage the unique benefits of logistics of project systems.

PROJECT CONSTRAINTS

Systems management is the pursuit of organizational goals within the constraints of time, cost, and quality expectations. The iron triangle model shows that project accomplishments are constrained by the boundaries of quality, time, and cost. In this case, quality represents the composite collection of project requirements. In a situation where precise optimization is not possible, there will have to be trade-offs between these three factors of success. The concept of iron triangle is that a rigid triangle of constraints encases the project. Everything must be accomplished within the boundaries of time, cost, and quality. If better quality is expected, a compromise along the axes of time and cost must be executed, thereby altering the shape of the triangle. The trade-off relationships are not linear and must be visualized in a multi-dimensional context. Scope requirements determine the project boundary and trade-offs must be done within that boundary. If we label the eight corners of the box as (a), (b), (c),..., (h), we can iteratively assess the best operating point for the project. For example, we can address the following two operational questions:

1. From the point of view of the project sponsor, which corner is the most desired operating point in terms of combination of requirements, time, and cost?
2. From the point of view of the project executor, which corner is the most desired operating point in terms of combination of requirements, time, and cost?

Note that all the corners represent extreme operating points. We notice that point (e) is the do-nothing state, where there are no requirements, no time allocation, and no cost incurrence. This cannot be the desired operating state of any organization that seeks to remain productive. Point (a) represents an extreme case of meeting all requirements with no investment of time or cost allocation. This is an unrealistic extreme in any practical environment. It represents a case of getting something for nothing. Yet, it is the most desired operating point for the project sponsor. By comparison, point (c) provides the maximum possible for requirements, cost, and time. In other words, the highest levels of requirements can be met if the maximum possible time is allowed and the highest possible budget is allocated. This is an unrealistic expectation in any resource-conscious organization. You cannot get everything you ask for to execute a project. Yet, it is the most desired operating point for the project executor. Considering the two extreme points of (a) and (c), it is obvious that the project must be executed within some compromise region within the scope boundary. A graphical analysis can reveal a possible view of a compromise surface with peaks and valleys representing give-and-take trade-off points within the constrained box. The challenge is to come up with some analytical modeling technique to guide decision-making over the compromise region. If we could collect sets of data over several repetitions of identical projects, then we could model a decision surface that can guide future executions of similar projects. Such typical repetitions of an identical project are most readily apparent in construction projects, for example residential home development projects.

Systems influence philosophy suggests the realization that you control the internal environment while only influencing the external environment. The inside (controllable) environment is represented as a black box in the typical input–process–output relationship. The outside (uncontrollable) environment is bounded by a cloud representation. In the comprehensive systems structure, inputs come from the global environment, are moderated by the immediate outside environment, and are delivered to the inside environment. In an unstructured inside environment, functions occur as blobs. A "blobby" environment is characterized by intractable activities where everyone is busy, but without a cohesive structure of input–output relationships. In such a case, the following disadvantages may be present:

- Lack of traceability
- Lack of process control
- Higher operating cost
- Inefficient personnel interfaces
- Unrealized technology potentials.

Organizations often inadvertently fall into the blobs structure because it is simple, low-cost, and less time-consuming, until a problem develops. A desired alternative is to model the project system using a systems value-stream structure. This uses a proactive and problem-preempting approach to execute projects. This alternative has the following advantages:

- Problem diagnosis is easier
- Accountability is higher
- Operating waste is minimized
- Conflict resolution is faster
- Value points are traceable.

DIVERSE APPLICATIONS OF DEJI SYSTEMS MODEL

To illustrate the diverse applications, versatility, and flexibility of the DEJI Systems Model, this section presents an unusual application of the systems approach to a familiar challenge in kitchen project management. This might appear to be an unconventional application, but it illustrates the fact that every pursuit is a project that is amenable to a systems methodology. By customizing the concepts and elements of Figure 1.1, we can achieve the project profile presented in Figure 1.2. The global framework of the model can be adapted for each specific application of interest. The implementation simply requires the inclusion of elements that pertain to the different topics of relevance in the prevailing context. This comprehensive framework ensures that all elements and activities that matter are embodied in the overall work plan for the domestic kitchen project. Of course, not all kitchen pursuits need to be as elaborate as illustrated in the figure. But the generic framework can serve as a reminder not to neglect critical items in the culinary pursuit. Similar to the diverse application in the kitchen, another popular application of the systems approach is in health care.

Emi Mahmoud, a Sudanese poet and activist, who won the 2015 Individual World Poetry Slam championship and later became a 2018 United Nations High Commissioner for Refugees (UNHCR) Goodwill Ambassador, is quoted as saying "My big passion in life is understanding different kinds of systems and how we break them down, and what they look like at their very core … and the most critical system is health care." She has the passion of applying a system approach to her aspired career diversification into the health care profession. Therein lies an affirmation of the efficacy of systems in every sphere of human pursuit. A systems framework works for all types of project management.

FIGURE 1.2   Adaptation of DEJI Systems Model for kitchen project management.

LEVERAGING ARTIFICIAL INTELLIGENCE FOR PROJECT MANAGEMENT

Project management has always taken advantage of whatever technology emerges from science, technology, engineering, and mathematics (STEM) research laboratories. The ongoing digital transformation in business and industry creates a new dimension of leveraging technological developments for improving project management. Specifically, artificial intelligence (AI) can aid project management in a variety of ways. The field of artificial intelligence continues to grow rapidly, fueled by the emergence of more advanced computers. The more computer power is available, the more AI can do for the benefit of project management. Many times, project management is about exploring options and making decisions quickly. This makes project management to be big-data-dependent. AI has the ability to handle and manipulate large amounts of data. The technique of "deep learning" that has developed in AI can enhance the reasoning process associated with project management. Thus, in the coming years, project management planning, organizing, scheduling, and control will leverage AI more and more. Consequently, project managers and analysts should be aware of the capabilities of AI and how to leverage it for project management functions. Project managers can benefit from knowing the basic origin of AI and what it entails. Badiru (1992), Badiru and Cheung (2002), and references therein, present a good historical evolution of AI, which is recounted in this section.

AI is defined as the intelligence exhibited by machines or software tools. It relates to the field of study which studies how to create computers and computer software that are capable of intelligent behavior. Intelligence is needed by project management. AI can provide additional "smarts" for better management of projects. Deep learning in AI systems is the ability of the system to use massive amounts of data and number crunching to self-generate intelligent actions, which is a desirable function in complex project management.

The background of AI has been characterized by controversial opinions and diverse approaches. Despite the controversies which have ranged from the basic definition of intelligence to questions about the moral and ethical aspects of pursuing AI, the technology continues to generate practical results. With increasing efforts in AI research, many of the prevailing arguments are being resolved with proven technical approaches. Expert systems (ES), the decision-making software branch of AI, are the most promising for project management applications.

AI is a controversial name for a technology that promises much potential for improving human productivity. The phrase seems to challenge human pride in being the sole creation capable of possessing real intelligence. Despite the doubters, AI is getting more and more into the mainstream of operations in business and industry. It is being shown again and again that AI may hold the key to improving operational effectiveness in many areas of applications. AI tools are being embedded in new software products, some of which are advertised as helping to solve problems that were, hitherto, unsolvable. Some observers have suggested changing the term "Artificial Intelligence" to a less-controversial one such as intelligent applications (IAs). This refers more to the way that computer and software are used innovatively to solve complex decision problems.

Natural intelligence involves the capability of humans to acquire knowledge, reason with the knowledge, and use it to solve problems effectively. By contrast, AI is defined as the ability of a machine to use simulated knowledge in solving problems.

The definition of intelligence had been sought by many great philosophers and mathematicians over the ages, including Aristotle, Plato, Copernicus, and Galileo. They attempted to explain the process of thought and understanding. The real key that started the quest for the simulation of intelligence did not occur, however, until the English philosopher Thomas Hobbes put forth an interesting concept in the 1650s. Hobbes believed that thinking consists of symbolic operations and that everything in life can be represented mathematically. These beliefs led directly to the notion that a machine capable of carrying out mathematical operations on symbols could imitate human thinking. This is the basic driving force behind the early AI movement. For that reason, Hobbes is sometimes referred to as the grandfather of AI.

While the term "artificial intelligence" was coined by John McCarthy relatively recently in1956, the idea had been considered centuries before. As long ago as 1637, Rene Descartes conceptually good job of imitating the human suintelligence when he said:

> For we can well imagine a machine so made that it utters words and even, in a few cases, words pertaining specifically to some actions that affect it physically. However, no such machine could ever arrange its words in various different ways so as to respond to the sense of whatever is said in its presence-as even the dullest people can do.

Descartes believed that the mind and the physical world are on parallel planes that cannot be equated. They are of different substances, following entirely different rules and can, thus, not be successfully compared. The physical world (i.e., machines) cannot imitate the mind because there is no common reference point.

The 1800s saw an advancement in the conceptualization of the computer. Charles Babbage, a British mathematician, laid the foundation for the construction of the computer, a machine defined as being capable of performing mathematical computations. Babbage introduced an analytical engine in 1833. This computational machine incorporated two unprecedented ideas that were to become crucial elements in the modem computer. First, it had operations that were fully programmable, and second, it could contain conditional branches. Without these two abilities, the power of today's computers would be inconceivable. Due to a lack of financial support, Babbage was never able to realize his dream of building the analytic engine. However, his dream was revived through the efforts of later researchers. Babbage's basic concepts can be observed in the way that most computers operate today.

Another British mathematician, George Boole, worked on issues that were to become equally important. Boole formulated the laws of thought that set up rules of logic for representing thought. The rules contained only two-valued variables. By this, any variable in a logical operation could be in one of only two states: yes or no, true or false, all or nothing, 0 or 1 (zero or one), on or off, and so on. This was the birth of digital logic, which formed a key component of the AI effort.

In the early 1900s, Alfred North Whitehead and Bertrand Russell extended Boole's logic to include mathematical operations. This not only led to the formulation of digital computers, but also made possible one of the first connections between computers and thought processes. However, there was still no acceptable way to construct such a computer.

In 1938, Claude Shannon demonstrated that Boolean logic consisting of only two-variable states (e.g., on–off switching of circuits) can be used to perform logic operations. Based on this premise, ENIAC (Electronic Numerical Integrator and Computer) was built in 1946 at the University of Pennsylvania. ENIAC was a large-scale, fully operational electronic computer that signaled the beginning of the first generation of digital computers. It could perform calculations 1,000 times faster than its electromechanical predecessors. It weighed 30 tons, stood two stories high, and occupied 1,500 square feet of floor space. Unlike today's computers, which operate in binary codes (0s and 1s), ENIAC operated in decimal (0, 1, 2,..., 9) and required ten vacuum tubes to represent one decimal digit. With over 18,000 vacuum tubes, ENIAC needed a great amount of electrical power, so much so that it was said that it dimmed the lights in Philadelphia whenever it operated.

Two of the leading mathematicians and computer enthusiasts between 1900 and 1950 were Alan Turing and John von Neumann. In 1945, von Neumann insisted that computers should not be built as glorified adding machines, with all their operations specified in advance. Rather, he suggested computers should be built as general-purpose logic machines capable of executing a wide variety of programs. Such machines, von Neumann proclaimed, would be highly flexible and capable of being readily shifted from one task to another. They could react intelligently to the results of their calculations, could choose among alternatives, and could even play checkers or chess. This represented something unheard of at that time: a machine with built-in intelligence, able to operate on internal instructions.

Prior to von Neumann's concept, even the most complex mechanical devices had always been controlled from the outside, for example, by setting dials and knobs. Von Neumann did not invent the computer, but what he introduced was equally significant: computing by use of computer programs, the way it is done today. His work paved the way for what would later be called artificial intelligence in computers.

Alan Turing also made major contributions to the conceptualization of a machine that can be universally used for all problems based only on variable instructions fed into it. Turing's universal machine concept, along with von Neumann's concept of a storage area containing multiple instructions that can be accessed in any sequence, solidified the ideas needed to develop the programmable computer. Thus, a machine was developed that could perform logical operations and could do them in varying orders by changing the set of instructions that were executed. Due to the fact that operational machines were now being realized, questions about the ·intelligence of the machines began to surface. Turin's other contribution to the world of AI came in the area of defining what constitutes intelligence. In 1950, he designed the Turing test for determining the intelligence of a system. The test utilized the conversational interaction between three players to try to verify computer intelligence.

The test is conducted by having a person (the interrogator) in a room that contains only a computer terminal. In an adjoining room, hidden from view, a man (person A) and a woman (person B) are located with another computer terminal. The interrogator communicates with the couple in the other room by typing questions on the keyboard. The questions appear on the couple's computer screen, and they respond by typing on their own keyboard. The interrogator can direct questions to either person A or person B, but without knowing which is the man and which is the woman.

The purpose of the test is to distinguish between the man and the woman merely by analyzing their responses. In the test, only one of the people is obligated to give truthful responses. The other person deliberately attempts to fool and confuse the interrogator by giving responses that may lead to an incorrect guess. The second stage of the test is to substitute a computer for one of the two persons in the other room. Now the human is obligated to give truthful responses to the interrogator, while the computer tries to fool the interrogator into thinking that it is human. Turing's contention is that if the interrogator's success rate in the human/computer version of the game is not better than his success rate in the man/woman version, then the computer can be said to be "thinking." That is, the computer possesses "intelligence." Turing's test has served as a classical example for AI proponents for many years.

By 1952, computer hardware had advanced far enough that actual experiments in writing programs to imitate thought processes could be conducted. The team of Herbert Simon, Allen Newell, and Cliff Shaw was organized to conduct such an experiment. They set out to establish what kinds of problems a computer could solve with the right programming. Proving theorems in symbolic logic, such as those set forth by Whitehead and Russell in the early 1900s, fits the concept of what they felt an intelligent computer should be able to handle.

It quickly became apparent that there was a need for a new higher-lever computer language than was currently available. First, they needed a language that was more user-friendly and could take program instructions that are easily understood by a human programmer and automatically convert them into machine language that could be understood by the computer. Second, they needed a programming language that changed the way in which computer memory was allocated. All previous languages would preassign memory at the start of a program. The team found that the type of programs they were writing would require large amounts of memory and would function unpredictably. To solve the problem, they developed a list processing language. This type of language would label each area of memory and then maintain a list of all available memory. As memory became available, it would update the list, and when more memory was needed, it would allocate the amount necessary. This type of programming also allowed the programmer to be able to structure his or her data so that any information that was to be used for a particular problem could be easily accessed. The end result of their effort was a program called Logic Theorist. This program had rules consisting of axioms already proved. When it was given a new logical expression, it would search through all of the possible operations in an effort to discover a proof of the new expression. Instead of using a brute force search method, they pioneered the use of heuristics in the search method.

The Logic Theorist that they developed in 1955 was capable of solving 38 of 52 theorems that Whitehead and Russell had devised. It did them very quickly. What took Logic Theorist a matter of minutes would have taken years if it had been done by simple brute force on a computer. By comparison, the steps that it went through to arrive at a proof to those that human subjects went through showed that it had achieved a remarkable imitation of the human thought process. This system is considered the first AI program.

The first AI conference happened in summer 1956, when the first attempt was made to establish the field of machine intelligence into an organized effort. The Dartmouth Summer Conference, organized by John McCarthy, Marvin Minsky, Nathaniel Rochester, and Claude Shannon, brought together people whose work and interest formally established the field of AI. The conference, held at Dartmouth College in New Hampshire, was funded by a grant from the Rockefeller Foundation. It was at that conference that John McCarthy coined the term "artificial intelligence." This was the same John McCarthy who developed the LISP programming language, which later became a standard tool for AI development. In fact, in those days some computer makers developed what they called LISP machines. This author did buy one of those machines for his AI lab at the University of Oklahoma in the early 1990s and developed courses in the applications of AI to project management and manufacturing (Badiru, 1996). In attendance at the first AI conference, in addition to the organizers, were Herbert Simon, Allen Newell, Arthur Samuel, Trenchard More, Oliver Selfridge, and Ray Solomonoff.

The Logic Theorist developed by Newell, Shaw, and Simon was discussed at the conference. Newell, Shaw, and Simon were far ahead of others in actually implementing AI ideas. The Dartmouth meeting served mostly as an avenue for the exchange of information and, more importantly, as a turning point in the main emphasis of work in the AI endeavor. Instead of concentrating on the hardware to imitate intelligence, the meeting set the course for examining the structure of the data being processed by computers, the use of computers to process symbols, the need for new languages, and the role of computers for testing theories.

The next major step in software technology came from Newell, Shaw, and Simon in 1959. The program they introduced was called General Problem Solver (GPS), which was intended to be a program that could solve many types of problems. It was capable of solving theorems, playing chess, or doing various complex puzzles. GPS was a significant step forward in AI. It incorporates several new ideas to facilitate problem-solving. The nucleus of the system was the use of means-end analysis, which involves comparing a present state with a goal state. The difference between the two states is determined, and a search is done to find a method to reduce this difference. This process is continued until there is no difference between the current state and the goal state.

In order to improve the search further, GPS contained two other features. The first is that, if while trying to reduce the deviation from the goal state, GPS finds that it has actually complicated the search process, it was capable of backtracking to an earlier state and exploring alternate solution paths. The second is that it was capable of defining sub-goal states that, if satisfied, would permit the solution process to continue. In formulating GPS, Newell and Simon had done extensive work studying human subjects and the way they solved problems. They felt that GPS did a good job of imitating the human subjects. They commented on the effort by saying:

> The fragmentary evidence we have obtained to date encourages us to think that the General Problem Solver provides a rather good first approximation to an information processing theory of certain kinds of thinking and problem-solving behavior. The processes of "thinking" can no longer be regarded as completely mysterious.

One criticism of GPS was that the only way the program obtained any information was through human input. The way and order in which the problems were presented was controlled by humans. Thus, the program was only doing what it was told to do. Newell and Simon argued that the fact that the program was not just repeating steps and sequences, but was actually applying rules to solve problems it had not previously encountered is indicative of intelligent behavior.

There were other criticisms as well. Humans are able to devise new shortcuts and improvise. GPS would always go down the same path to solve the same problem, making the same mistakes as before. It could not learn. In solving problems, it was difficult to determine what search space to use. Sometimes, solving the problem is trivial compared to finding the search space. The problems posed to GPS were all of a specific nature. They were all puzzles or logical challenges, problems that could easily be expressed in symbolic form and operated on in a pseudo-mathematical approach. There are many problems that humans face that are not so easily expressed in a symbolic form.

Also, in 1959, John McCarthy came out with a tool that was to greatly improve the ability of researchers to develop AI programs. He developed a new computer programming language called LISP (list processing). It was to become one of the most widely used languages in the field.

LISP is distinctive in two areas: memory organization and control structure. The memory organization is done in a tree fashion with interconnections between memory groups. Thus, it permits a programmer to keep track of complex structural relationships. The other distinction is the way the control of the program is done. Instead of working from the prerequisites to a goal, it starts with the goal and works backwards to determine what prerequisites are required to achieve the goal.

In 1960, Frank Rosenblatt did work in the area of pattern recognition. He introduced a device called PERCEPTRON that was supposed to be capable of recognizing letters and other patterns. It consisted of a grid of 400 photo cells connected with wires to a response unit that would produce a signal only if the light coming off the subject to be recognized crossed a certain threshold.

During the latter part of the 1960s, there were two efforts in another area of simulating human reasoning. Kenneth Colby at Stanford University and Joseph Weizenbaum at MIT wrote separate programs that were capable of interacting in a two-way conversation. Weizenbaum's program was called ELIZA. The programs were able to sustain very realistic conversations by using very clever techniques. For example, ELIZA used a pattern-matching method that would scan for keywords like "I," "you," and "like,". If one of these words was found, it would execute rules associated with it. If no match was found, the program would respond with a request for more information or with a noncommittal response.

It was also during the 1960s that Marvin Minsky and his students at MIT made significant contributions toward the progress of AI. One student, T. G. Evans, wrote a program that would perform visual analogies. The program was shown two figures that had some relationship to each other and was then asked to find another set of figures from a set that matched the same relationship. The input to the computer was not done by a visual sensor (like the one worked on by Rosenblatt), but instead the figures were described to the system.

In 1968, another student of Minsky, Daniel Bobrow, came out with a linguistic problem-solver called STUDENT. It was designed to solve problems that were presented to it in a word-problem format. The key to the program was the assumption that every sentence was an equation. It would take certain words and turn them into mathematical operations. For example, it would convert "is" into "=" and "per" into "/."

Even though STUDENT responded very much the same way that a real student would, there was a·major difference in the depth of understanding. While the program was capable of calculating the time two trains would collide given the starting points and speeds of both, it had no real understanding or even cared what a "train" or "time" was. Expressions like "perchance" and "this is it" could mean totally different things than what the program would assume. A human student would be able to discern the intended meaning from the context in which the terms were used.

In an attempt to answer the criticisms about understanding, another student at MIT, Terry Winograd, developed a significant program named SHRDLU. In setting up his program, he utilized what was referred to as a micro-world or blocks-world. This limited the scope of the world that the program had to try to understand. The program communicated in what appeared to be natural language.

The world of SHRDLU consisted of a set of blocks of varying shapes (cubes, pyramids, etc.), sizes, and colors. These blocks were all set on an imaginary table. Upon request, SHRDLU would rearrange the blocks to any requested configuration. The program was capable of knowing when a request was unclear or impossible. For instance, if it was requested to put a block on top of a pyramid, it would request that the user specify more clearly what block and what pyramid. It would also recognize that the block would not sit on top of the pyramid. Two other approaches that the program took that were new to programs were the ability to make assumptions and the ability to learn. If asked to pick up a larger block, it would assume that you meant a larger block than the one it was currently working on. If asked to build a figure that it did not know, it would ask for an explanation of what it was and, thereafter, it would recognize the object. One major sophistication that SHRDLU added to the science of AI programming was its use of a series of expert modules or specialists. There was one segment of the program that specialized in segmenting sentences into meaningful word groups, a sentence specialist to determine the relationship between nouns and verbs, and a scenario specialist that understood how individual scenes related to one another. This sophistication greatly enhanced the method in which instructions were analyzed.

As sophisticated as SHRDLU was at that time, other scholars were quick to point out its deficiencies. SHRDLU only responded to requests; it could not initiate conversations. It also had no sense of conversational flow. It would jump from performing one type of task to a totally different one if so requested. While SHRDLU had an understanding of the tasks it was to perform and the physical world in which it operated, it still could not understand very abstract concepts.

The various attempts at formally defining the use of machines to simulate human intelligence led to the development of several branches of AI. Some of the sub-specialties of AI include

1. Natural language processing deals with various areas of research such as database inquiry systems, story understanders, automatic text indexing, grammar and style analysis of text, automatic text generation, machine translation, speech analysis, and speech synthesis.
2. Computer Vision deals with research efforts involving scene analysis, image understanding, and motion derivation.

3. Robotics involves the control of effectors on robots to manipulate or grasp objects, locomotion of independent machines, and use of sensory input to guide actions.
4. Problem-solving and planning involve applications such as refinement of high-level goals into lower-level ones, determination of actions needed to achieve goals, revision of plans based on intermediate results, and focused search of important goals.
5. Learning deals with research into various forms of learning including rote learning, learning through advice, learning by example, learning by task performance, and learning by following concepts. In recent years, deep learning has become a key approach to getting AI to be more robust in independent thinking.
6. Expert systems deal with the processing of knowledge as opposed to the processing of data. It involves the development of computer software to solve complex decision problems.

Neural networks, sometimes called connectionist systems, are networks of simple processing elements or nodes capable of processing information in response to external inputs. Neural networks were originally presented as models of the human nervous system. Just after World War II, scientists found out that the physiology of the brain was similar to the electronic processing mode used by computers. In both cases, large amounts of data are manipulated. In the case of computers, the elementary unit of processing is the bit, which is in either an "on" or "off" state. In the case of the brain, neurons perform the basic data processing. Neurons are tiny cells that follow a binary principle of being either in a state of firing (on) or not firing (off). When a neuron is on, it fires a signal to other neurons across a network of synapses. In the late 1940s, Donald Hebb, a researcher, hypothesized that biological memory results when two neurons are active simultaneously. The synaptic connection of synchronous neurons is reinforced and given preference over connections made by neurons that are not active simultaneously. The level of preference is measured as a weighted value. Pattern recognition, a major strength of human intelligence, is based on the weighted strengths of the reinforced connections between various pairs of simultaneously active neurons. The idea presented by Hebb was to develop a computer model based on the way in which neurons form connections in the human brain. But the idea was considered to be preposterous at that time since the human brain contains 100 billion neurons and each neuron is connected to 10,000 others by a synapse, unless it is a neuron from a moron. Even with today's computing capability, it is still difficult to duplicate the activities of neurons.

In 1969, Marvin Minsky and Seymour Pappert criticized existing neural network research as being worthless. It has been claimed that the pessimistic views they presented discouraged further funding for neural network research for several years. Funding was diverted instead to further research of expert systems, which Minsky and Pappert favored. Fortunately, neural networks made a strong comeback later on. Because neural networks are modeled after the operations of the brain, they hold considerable promise as building blocks for achieving the ultimate aim of AI. The present generation of neural networks uses artificial neurons. Each neuron is connected to at least one other neuron in a synapse-like fashion. The networks are based on some form of learning model. Neural networks learn by evaluating changes in input. Learning can be either supervised or unsupervised. In supervised learning, each response is guided by given parameters. The computer is instructed to compare any inputs to ideal responses, and any discrepancy between the new inputs and ideal responses is recorded. The system then uses this data bank to guess how much the newly gathered data are similar to or different from the ideal responses. That is, how closely the pattern matches. Supervised learning networks are now commercially used for control systems, handwriting analysis, speech recognition, and other science and technology applications.

In unsupervised learning, input is evaluated independently and stored as a pattern. The system evaluates a range of patterns, and identifies similarities and dissimilarities among them. However, the system cannot derive any meaning from the information without human assignment of values to the patterns. Comparisons are relative to other results, rather than to an ideal result. Unsupervised learning networks are used to discover patterns where a particular outcome is not known in advance, such as in physics research and the analysis of financial data. Several commercial neural network products have entered the market over the years, including NeuroShell from Ward Systems Group. The software is expensive but is relatively easy to use. It interfaces well with other software such as spreadsheets, as well as with C and other programming languages.

Despite the proven potential of neural networks, they drastically oversimplify the operations of the brain. The existing systems can only undertake elementary pattern-recognition tasks and are weak at deductive reasoning, math calculations, and other computations that are easily handled by conventional computer processing. The difficulty in achieving the promise of neural networks lies in our limited understanding of how the human brain functions. Undoubtedly, to model the brain accurately, we must know more about it. But a complete knowledge of the brain is still many years away.

From the late 1960s to the early 1970s, a special branch of AI began to emerge. The branch, known as expert systems, grew dramatically due to its ability to deliver results, albeit not in a thinking context. For years, expert systems represented the most successful demonstration of the capabilities of AI. Expert systems were the first truly commercial applications of the work done in the AI field and, as such, received considerable publicity for many years.

Unlike the desire to develop general problem-solving techniques that had characterized AI before, expert systems address problems that are focused. When Edward Feigenbaum developed the first successful expert system, DENDRAL, he had a specific type of problem that he wanted to be able to solve. The problem involved determining which organic compound was being analyzed in a mass spectrograph. The program was intended to simulate the work that an expert chemist would do in analyzing the data. This led to the term "expert system."

Between 1970 and 1980, numerous expert systems were introduced to handle several functions, from diagnosing diseases to analyzing geological exploration information. Of course, expert systems have not escaped the critics. Due to the nature of the system, critics argue that it does not fit the true structure of AI. Because of the use of only specific knowledge and the ability to solve only specific problems, some critics are apprehensive about referring to an expert system as intelligent. Proponents argue that if the system produces the desired results, it is of little concern whether it is intelligent or not.

In 1972, Hubert Dreyfus initiated another debate of interest. Joseph Weizenbaum presented similar views in 1976. The issues that both authors raised touched on some of the basic questions that dated back to the time of Descartes. One of Weizenbaum's reservations concerned what should ethically and morally be handed over to machines. He maintained that the path that AI was pursuing was headed in a dangerous direction. Some aspects of human experience, such as love and morality, cannot be adequately imitated by machines.

While the debates were going on over how much AI could do, the work on getting AI to do more continued. In 1972, Roger Shrank introduced the notion of script, the set of familiar events that can be expected from a frequently encountered setting. This enables a program to assimilate facts quickly. In 1975, Marvin Minsky presented the idea of frames. Even though neither concept drastically advanced the theory of AI, they did help expedite research in the field.

In 1979, Minsky suggested a method that could lead to a better simulation of intelligence: the society of minds view, in which the execution of knowledge is performed by several programs working in conjunction simultaneously. This concept helped to encourage interesting developments such as present-day parallel processing.

During the 1980s, AI gained significant exposure and interest. AI, once restricted to the domain of esoteric research, has now become a practical tool for solving real problems. While AI is enjoying its most prosperous period, it is still plagued with disagreements, criticism, and skepticism. The emergence of commercial expert systems on the market has created both enthusiasm and skepticism. There is no doubt that more research and successful applications developments will help prove the potential of expert systems. It should be recalled that new technologies sometimes fail to convince all initial observers. IBM, which later became a giant in the personal computer business, hesitated for several years before getting into the microcomputer market because the company never thought that those little boxes called personal computers would ever have any significant impact on the society. Today's laptop market has proven otherwise.

Nowadays, more expert systems are showing up, not as stand-alone systems, but as software tools (or apps) embedded in large software systems. This trend is bound to continue as systems integration takes hold in many software applications. Many conventional commercial packages, such as statistical analysis systems, data management systems, information management systems, project management systems, and data analysis systems, now contain embedded heuristics that constitute expert systems components of the packages. Even some computer operating systems now contain embedded expert systems designed to provide real-time systems monitoring and troubleshooting. With the success of embedded expert systems, the long-awaited payoffs from AI systems may be realized on a larger scope in the coming years. Because the technology behind expert systems has changed little over the past years, the issue is not whether the technology is useful, but how to implement it more fruitfully. This is why the integrated systems approach of this book is very useful. This book focuses not only on the traditional principles of project management, but also on the emerging technological tools that can aid project planning, scheduling, and control. Combining neural networks with expert systems, for example, will become more prevalent in the modern AI applications. In combination, the neural networks might be implemented as a tool for scanning and selecting data, while the expert system would evaluate the data and present recommendations. The effort in AI is worthwhile as long as it increases the understanding that we have of intelligence and enables us to do things that we previously could not do.

HISTORICAL CONTEXT OF AI APPLICATIONS TO PROJECT MANAGEMENT

Due to the discoveries made in AI and ES research, computers are now capable of things that were once beyond imagination. The field of project management will be hugely impacted as AI systems become more robust and more intelligent. In the early 1980s, the field of industrial engineering actively embraced the application of desktop microcomputers, which later evolved to laptop computers, for project management and other decision-making applications (Badiru and Whitehouse, 1989). The author and his graduate students developed several such applications as evidenced the by the selected list below:

- Development of An Intelligent Transportation Network Model for Complex Economic and Infrastructure Simulations
- An Intelligent Computer Modeling Approach for Critical Resource Diagramming Network Analysis in Project Scheduling
- Intelligent Implementation of Critical Path Method and Critical Resource Diagramming Using Arena Simulation Software
- Development of Mode Estimating Technique for Cantor Search of Sorted Data in Artificial Intelligence Systems and Manufacturing Design
- Expert System Simulation Metamodel for Shop Floor Performance Analysis
- An Expert System Model for Supplier Development Program
- A Prototype Expert System for Integrated Project Management
- An Integrated Expert System with a Fuzzy Linguistic Model for Facilities Layout
- A Methodology for A Real-Time Artificial Intelligence Surveillance System
- An Expert System Model for Pareto Analysis
- PC Opal: AI-based PC Tool for Optimizing Parameter Levels: An Expert System to Design Experiments Using Taguchi's Orthogonal Arrays and to Analyze Responses
- ROBEX (Robot Expert): An Expert System for Manufacturing Robot System Implementation

- JUSTEX: An Expert System for the Justification of Advanced Manufacturing Technology
- An Expert System for Simulation Modeling of Project Networks
- PROCESS-PLUS: A Prototype Expert System for Generative Process Planning
- ROBCON: A Prototype Expert System for Robot Consultancy
- Expert System for Heuristic Selection for Project Scheduling
- In 1982, Gary Whitehouse, assisted by Adedeji Badiru, organized the first Computers and Industrial Engineering Conference in Orlando, Florida, the proceedings of which later morphed into the archival refereed **Computers and Industrial Engineering Journal**, through which new intelligent applications of computer tools have been published consistently over the years.

PROJECT WORKFORCE DEVELOPMENT IN A COVID-19 ENVIRONMENT

Project goals and objectives are accomplished through human resources, working in consonant with technology assets. According to Badiru and Barlow (2020), the unfolding workforce disruption caused by COVID-19 has necessitated a new focus on the challenges of workforce development in the State of Ohio. The decimation of productivity caused by COVID-19 requires not only the traditional strategies of workforce development, but also the uncharted territory of workforce redevelopment and preservation. Reporting during COVID-19 indicates a precipitous decline in the ability of the workforce to continue to contribute to economic development and vitality of the State during the lockdown. When businesses open again, it will be necessary for workers to relearn their jobs to return to the level of proficiency and efficiency needed to move the State's economy forward. The technical topic of learning curve analysis postulates that performance improves with repeated cycles of operations. Whenever work performance is interrupted for a prolonged period, as we are currently experiencing, the processes of natural forgetting or technical regressing set in. To offset this decline, direct concerted efforts must be made beyond anything we have experienced before. This urgency to recover the economy led to our call for new innovations in workforce development and redevelopment. We cannot be lackadaisical in leaving things to the normal process of regaining form, routine, and function.

Typically, we erroneously focus on technical tools as the embodiment of innovation. But more often than not, process innovations might be just as vital. Workforce development, in particular, is more process development than tool development. There are numerous human factors strategies that can enhance the outcomes of workforce development. Some of the innovations we recommend in this regard include paying attention to the hierarchy of needs of the worker (primarily safety in our current world), recognizing the benefits of diversity, elevating the visibility of equity, instituting efforts to negate adverse aspects of cultural bias, and appreciating the dichotomy of socio-economic infrastructure. While not too expensive to implement, these innovative strategies can be tremendously effective. Workforce redevelopment is a topic not very often discussed, but COVID-19 brings its importance to the forefront. Redevelopment will be needed not only to boost the quantity of the productive capacity, but also to restore and augment the capability, availability, and reliability of the workforce beyond the previous performance yardstick.

The greatest challenge in a COVID-19 environment is workforce preservation. We don't think there will be a post-COVID-19 environment in the near future. Coronavirus, the virus that causes COVID-19, is something we may have to contend with cyclically into the foreseeable future. How do we preserve the workforce in such a persistent COVID environment? Preservation of a well-developed workforce can only be assured through innovative health and safety safeguards, as well as new organizational processes and procedures that will take a thorough understanding of the recurring risks that may be posed by virus outbreaks. A workforce member who becomes ill or decides to leave an organization is a workforce member that we fail to preserve. Typically, a society addresses safety and security as necessary social mandates. Our postulation here is that we need to elevate that perception to the level of workforce necessity.

A workforce that is well educated and well developed, but stymied by the implications of a virus cannot be a productive workforce that contributes to the continued economic development of a Region or State. Institutions of higher learning, such as AFIT, and state-level workforce development organizations, such as SOCHE (Southwest Ohio Council for Higher Education), continue to partner in addressing new innovations in workforce development, redevelopment, and preservation. More widespread partnerships are needed in this effort that bodes well for the economic health and vitality of the community and any project environment.

CONCLUSIONS

As has been presented in this chapter, a systems approach increases the potential for success for any project, whether small or large, whether public or private, whether profit-oriented or non-profit-oriented, whether corporate or personal, and whether technical or non-technical. In each case, a structured and systems-guided framework is desired. This is the efficacy of the DEJI Systems Model. The model facilitates continuity, consistency, stability, uniformity, trust, confidence, credibility, integrity, and a sense of decency in project execution. Further, it can preempt the potential for a loss of professional dignity that sometimes plaque large multinational projects. This chapter concludes with a community-focused commentary on systems-based innovations in workforce development, redevelopment, and preservation.

REFERENCES

Badiru, A. B. (1992). *Expert systems applications in engineering and manufacturing.* New Jersey: Prentice-Hall.

Badiru, A. B. (1996). *Project management in manufacturing and high technology operations* (2nd ed.). New York: John Wiley & Sons.

Badiru, A. B. (2019). *Project management: systems, principles, and applications* (2nd ed.). Boca Raton, FL: Taylor & Francis Group/CRC Press.

Badiru, A. & Barlow, C. (2020, May 15). Developing workforce in era of COVID-19. *Dayton Daily News newspaper*, p. B7.

Badiru, A. B. & Cheung, J. (2002). *Fuzzy engineering expert systems with neural network applications.* New York: John Wiley & Sons.

Badiru, A. B. & Whitehouse, Gary E. (1989). *Computer tools, models, and techniques for project management.* Blue Ridge Summit, PA: TAB Professional & Reference Books, Inc.

Project Design

2

The project that you have the right to do is
not always the right project to do
– Adedeji Badiru

DEFINITION OF PROJECT DESIGN

Design, in the context of project management, is not limited to physical or technological design of a product. For the purpose of project management, as embodied in the DEJI Systems Model, design can define a project concept, a project organization structure, a project breakdown structure, and other pertinent project characteristics. The DEJI Systems Model was originally developed for product design and development processes. The applications of the technique have since been expanded to general applications, whereby the definition of design is more flexible. This chapter covers the various interpretations of design to represent diverse project management needs. Systems view of a project makes the project execution more adaptive, agile, efficient, and effective. A system is a collection of interrelated elements working together synergistically to achieve a set of objectives. Any project is, in effect, a collection of interrelated activities, people, tools, resources, processes, and other assets brought together in the pursuit of a common goal. The project goal may be in terms of the following alternate outputs:

- A physical product
- A desired service
- A specific result.

This makes it possible to view any project as a system that is amenable to all the classical and modern concepts of systems management. The DEJI Systems Model, as a nascent methodology of systems management, has proven effective for harnessing the conventional tools and techniques of project management. Project design should provide insight and guidance for project scoping, planning, organizing, scheduling, resource allocation, tracking, reporting, control, and phaseout. Not all conceptual designs are worthy of becoming a formal project. Not all projects that are justified are desirable. Not all desirable projects are practical in implementation. Not all implemented projects are acceptable to stakeholders.

Organizational performance is predicated on a multitude of factors, some are quantitative, while some are qualitative. Systems engineering efficiency and effectiveness are of interest across the spectrum of the diversity of organizational performance under the platform of project management. Project analysts should be interested in having systems engineering serve as the umbrella for improvement efforts throughout the organization. This will get everyone properly connected with the prevailing organizational goals as well as create collaborative avenues among the personnel. Systems application applies across the spectrum of any organization and encompasses the following elements:

- Technological systems (e.g., engineering control systems and mechanical systems)
- Organizational systems (e.g., work process design and operating structures)
- Human systems (e.g., interpersonal relationships and human–machine interfaces).

For example, the pursuit of organizational or enterprise transformation is best achieved through the involvement of everyone, from a systems perspective. Every project environment is very complex because of the diversity of factors involved. There are differing human personalities. There are differing technical requirements. There are differing expectations. There are differing environmental factors. Each specific context and prevailing circumstances determine the specific flavor of what can and cannot be done in the project. The best approach for effective project management is to design the project to adapt to what each project requires.

Project management continues to grow as an effective means of managing functions in any organization. Project management should be an enterprise-wide systems-based endeavor. Enterprise-wide project management is the application of project management techniques and practices across the full scope of the enterprise. This concept is also referred to as management by project (MBP). MBP is a contemporary concept that employs project management techniques in various functions within an organization.

MBP recommends pursuing endeavors as project-oriented activities. It is an effective way to conduct any business activity. It represents a disciplined approach that defines any work assignment as a project. Under MBP, every undertaking is viewed as a project that must be managed just like a traditional project. The characteristics required of each project so defined are

1. An identified scope and a goal
2. A desired completion time
3. Availability of resources
4. A defined performance measure
5. A measurement scale for review of work.

An MBP approach to operations helps in identifying unique entities within functional requirements. This identification helps determine where functions overlap and how they are interrelated, thus paving the way for better planning, scheduling, and control. Enterprise-wide project management facilitates a unified view of organizational goals and provides a way for project teams to use information generated by other departments to carry out their functions.

The use of project management continues to grow rapidly. The need to develop effective management tools increases with the increasing complexity of new technologies and processes. The life cycle of a new product to be introduced into a competitive market is a good example of a complex process that must be managed with integrative project management approaches. The product will encounter management functions as it goes from one stage to the next. Project management will be needed throughout the design and production stages of the product. Project management will be needed in developing marketing, transportation, and delivery strategies for the product. When the product finally gets to the customer, project management will be needed to integrate its use with those of other products within the customer's organization.

The need for a project management approach is established by the fact that a project will always tend to increase in size even if its scope is narrowing. The following three literary laws are applicable to any project environment:

Parkinson's law: Work expands to fill the available time or space.
Peter's principle: People rise to the level of their incompetence.
Murphy's law: Whatever can go wrong will.
Badiru's rule: The grass is always greener where you most need it to be dead.

A systems-based design of the project can help diminish the adverse impacts of these laws through good project planning, organizing, scheduling, and control.

TOOLS FOR PROJECT DESIGN

Project management tools can be classified into three major categories:

1. *Qualitative tools*: There are the managerial tools that aid in the interpersonal and organizational processes required for project management.
2. *Quantitative tools*: These are analytical techniques that aid in the computational aspects of project management.
3. *Computer tools*: These are software and hardware tools that simplify the process of planning, organizing, scheduling, and controlling a project. Software tools can help in both the qualitative and quantitative analyses needed for project management.

No matter how good a technology is, and no matter how enhanced a process might be, it is ultimately the people involved that determine success. This makes it imperative to take care of people issues first in the overall systems approach to project management. Many organizations recognize this, but only few have been able to actualize the ideals of managing people productively. Execution of operational strategies requires forthrightness, openness, and commitment to get things done. Lip service and arm waving are not sufficient. Tangible programs that cater to the needs of people must be implemented. It is essential to provide incentives, encouragement, and empowerment for people to be self-actuating in determining how best to accomplish their job functions. A summary of critical factors for systems success encompasses the following:

Total system management (hardware, software, and people)
Operational effectiveness
Operational efficiency
System suitability
System resilience
System affordability

System supportability
System life cycle cost
System performance
System schedule
System cost.

BODY OF KNOWLEDGE FRAMEWORK

The general body of knowledge (PMBOK®) for project management is published and disseminated by the Project Management Institute (PMI). The body of knowledge comprises specific knowledge areas, which are organized into the following broad areas:

1. Project *integration* management
2. Project *scope* management
3. Project *time* management
4. Project *cost* management
5. Project *quality* management
6. Project *human resource* management
7. Project *communications* management
8. Project *risk* management
9. Project *procurement and subcontract* management.

The listed segments of the body of knowledge of project management cover the range of functions associated with any project, particularly complex ones. Multinational projects particularly pose unique challenges pertaining to reliable power supply, efficient communication systems, credible government support, dependable procurement processes, consistent availability of technology, progressive industrial climate, trustworthy risk mitigation infrastructure, regular supply of skilled labor, uniform focus on quality of work, global consciousness, hassle-free bureaucratic processes, coherent safety and security system, steady law and order, unflinching focus on customer satisfaction, and fair labor relations. Assessing and resolving concerns about these issues in a step-by-step fashion will create a foundation of success for a large project. While no system can be perfect and satisfactory in all aspects, a tolerable trade-off on the factors is essential for project success.

THE KNOWLEDGE AREAS

The key components of each element of the body of knowledge are summarized as follows:

Integration management
   Integrative project charter
   Project scope statement
   Project management plan
   Project execution management
   Change control

Scope management
   Focused scope statements
   Cost/benefits analysis
   Project constraints
   Work breakdown structure
   Responsibility breakdown structure
   Change control

Time management
   Schedule planning and control
   PERT and Gantt charts
   Critical path method
   Network models
   Resource loading
   Reporting

Cost management
   Financial analysis
   Cost estimating
   Forecasting
   Cost control
   Cost reporting

Quality management
   Total quality management
   Quality assurance
   Quality control
   Cost of quality
   Quality conformance

Human resource management
  Leadership skill development
  Team building
  Motivation
  Conflict management
  Compensation
  Organizational structures

Communications management
  Communication matrix
  Communication vehicles
  Listening and presenting skills
  Communication barriers and facilitators

Risk management
  Risk identification
  Risk analysis
  Risk mitigation
  Contingency planning

Procurement and subcontract management
  Material selection
  Vendor prequalification
  Contract types
  Contract risk assessment
  Contract negotiation
  Contract change orders

STEP-BY-STEP AND COMPONENT-BY-COMPONENT IMPLEMENTATION

The efficacy of the systems approach is based on step-by-step and component-by-component implementation of the project management process. The major knowledge areas of project management are administered in a structured outline covering six basic clusters consisting of the following:

1. Initiating
2. Planning
3. Executing

4. Monitoring
5. Controlling
6. Closing.

The implementation clusters represent five process groups that are followed throughout the project life cycle. Each cluster itself consists of several functions and operational steps. When the clusters are overlaid on the nine knowledge areas, we obtain a two-dimensional matrix that spans 44 major process steps. The monitoring and controlling clusters are usually administered as one lumped process group (monitoring and controlling). In some cases, it may be helpful to separate them to highlight the essential attributes of each cluster of functions over the project life cycle. In practice, the processes and clusters do overlap. Thus, there is no crisp demarcation of when and where one process ends and where another one begins over the project life cycle. In general, project life cycle defines the following:

1. Resources that will be needed in each phase of the project life cycle
2. Specific work to be accomplished in each phase of the project life cycle.

The project design covers major phases of project life cycle going from the conceptual phase through the close-out phase. It should be noted that project life cycle is distinguished from product life cycle. Project life cycle does not explicitly address operational issues, whereas product life cycle is mostly about operational issues starting from the product's delivery to the end of its useful life. Note that for technical projects, the shape of the life cycle curve may be expedited due to the rapid developments that often occur in science, technology, and engineering activities. For example, for a high-technology project, the entire life cycle may be shortened, with a very rapid initial phase, even though the conceptualization stage may be very long. Typical characteristics of project life cycle include the following:

1. Cost and staffing requirements are lowest at the beginning of the project and ramp up during the initial and development stages.
2. The probability of successfully completing the project is lowest at the beginning and highest at the end. This is because many unknowns (risks and uncertainties) exist at the beginning of the project. As the project nears its end, there are fewer opportunities for risks and uncertainties.
3. The risks to the project organization (project owner) are lowest at the beginning and highest at the end. This is because not much investment has gone into the project at the beginning, whereas

much has been committed by the end of the project. There is a higher sunk cost manifested at the end of the project.

4. The ability of the stakeholders to influence the final project outcome (cost, quality, and schedule) is highest at the beginning and gets progressively lower toward the end of the project. This is intuitive because influence is best exerted at the beginning of an endeavor.

5. Value of scope changes decreases over time during the project life cycle, while the cost of scope changes increases over time. The suggestion is to decide and finalize scope as early as possible. If there are to be scope changes, do them as early as possible.

PROJECT DESIGN STRUCTURE

The overall project management systems execution can be outlined as summarized in the following.

Problem Identification

Problem identification is the stage where a need for a proposed project is identified, defined, and justified. A project may be concerned with the development of new products, implementation of new processes, or improvement of existing facilities.

Project Definition

Project definition is the phase at which the purpose of the project is clarified. A *mission statement* is the major output of this stage. For example, a prevailing low level of productivity may indicate a need for a new manufacturing technology. In general, the definition should specify how project management may be used to avoid missed deadlines, poor scheduling, inadequate resource allocation, lack of coordination, poor quality, and conflicting priorities.

Project Planning

A plan represents the outline of the series of actions needed to accomplish a goal. Project planning determines how to initiate a project and execute its objectives. It may be a simple statement of a project goal, or it may be a detailed account of procedures to be followed during the project. Planning can be summarized as follows:

Objectives
Project definition
Team organization
Performance criteria (time, cost, and quality)

Project Organizing

Project organization specifies how to integrate the functions of the personnel involved in a project. Organizing is usually done concurrently with project planning. Directing is an important aspect of project organization. Directing involves guiding and supervising the project personnel. It is a crucial aspect of the management function. Directing requires skillful managers who can interact with subordinates effectively through good communication and motivation techniques. A good project manager will facilitate project success by directing his or her staff, through proper task assignments, toward the project goal.

Workers perform better when there are clearly defined expectations. They need to know how their job functions contribute to the overall goals of the project. Workers should be given some flexibility for self-direction in performing their functions. Individual worker needs and limitations should be recognized by the manager when directing project functions. Directing a project requires skills dealing with motivating, supervising, and delegating.

Resource Allocation

Project goals and objectives are accomplished by allocating resources to functional requirements. Resources can consist of money, people, equipment, tools, facilities, information, skills, and so on. These are usually in short supply. The people needed for a particular task may be committed to other ongoing projects. A crucial piece of equipment may be under the control of another team. Chapter 5 addresses resource allocation in detail.

Project Scheduling

Timeliness is the essence of project management. Scheduling is often the major focus in project management. The main purpose of scheduling is to allocate resources so that the overall project objectives are achieved within a reasonable time span. Project objectives are generally conflicting in nature. For example, minimization of the project completion time and minimization of the project cost are conflicting objectives. That is, one objective is improved at the expense of worsening the other objective. Therefore, project scheduling is a multiple-objective decision-making problem.

In general, scheduling involves the assignment of time periods to specific tasks within the work schedule. Resource availability, time limitations, urgency level, required performance level, precedence requirements, work priorities, technical constraints, and other factors complicate the scheduling process. Thus, the assignment of a time slot to a task does not necessarily ensure that the task will be performed satisfactorily in accordance with the schedule. Consequently, careful control must be developed and maintained throughout the project scheduling process.

Project Tracking and Reporting

This phase involves checking whether or not project results conform to project plans and performance specifications. Tracking and reporting are prerequisites for project control. A properly organized report of the project status will help identify any deficiencies in the progress of the project and help pinpoint corrective actions.

Project Control

Project control requires that appropriate actions be taken to correct unacceptable deviations from expected performance. Control is actuated through measurement, evaluation, and corrective action. Measurement is the process of measuring the relationship between planned performance and actual performance with respect to project objectives. The variables to be measured, the measurement scales, and the measuring approaches should be clearly specified during the planning stage. Corrective actions may involve rescheduling, reallocation of resources, or expedition of task performance. Project control involves the following:

Tracking and reporting
Measurement and evaluation
Corrective action (plan revision, rescheduling, and updating).

Project Termination

Termination is the last stage of a project. The phaseout of a project is as important as its initiation. The termination of a project should be implemented expeditiously. A project should not be allowed to drag on after the expected completion time. A terminal activity should be defined for a project during the planning phase. An example of a terminal activity may be the submission of a final report, the power on of new equipment, or the signing of a release order. The conclusion of such an activity should be viewed as the completion of the project. Arrangements may be made for follow-up activities that may improve or extend the outcome of the project. These follow-up or spin-off projects should be managed as new projects but with proper input–output relationships within the sequence of projects.

Project design under the DEJI Systems Model still encompasses the traditional project management framework encompassing the broad sequence of the following categories:

Planning → Organizing → Scheduling → Control → Termination

An outline of the functions to be carried out during a project should be made during the planning stage of the project. A model for such an outline is presented hereafter. It may be necessary to rearrange the contents of the outline to fit the specific needs of a project.

Planning

1. Specify project background
   a. Define current situation and process
      i. Understand the process
      ii. Identify important variables
      iii. Quantify variables
   b. Identify areas for improvement
      i. List and discuss the areas
      ii. Study potential strategy for solution
2. Define unique terminologies relevant to the project
   a. Industry-specific terminologies
   b. Company-specific terminologies
   c. Project-specific terminologies

3. Define project goal and objectives
   a. Write mission statement
   b. Solicit inputs and ideas from personnel

4. Establish performance standards
   a. Schedule
   b. Performance
   c. Cost
5. Conduct formal project feasibility study
   a. Determine impact on cost
   b. Determine impact on organization
   c. Determine project deliverables
6. Secure management support

Organizing

1. Identify project management team
   a. Specify project organization structure
      i. Matrix structure
      ii. Formal and informal structures
      iii. Justify structure
   b. Specify departments involved and key personnel
      i. Purchasing
      ii. Materials management
      iii. Engineering, design, manufacturing, and so on
   c. Define project management responsibilities
      i. Select project manager
      ii. Write project charter
      iii. Establish project policies and procedures
2. Implement triple C model
   a. Communication
      i. Determine communication interfaces
      ii. Develop communication matrix
   b. Cooperation
      i. Outline cooperation requirements, policies, and procedures
   c. Coordination
      i. Develop work breakdown structure
      ii. Assign task responsibilities
      iii. Develop responsibility chart

Scheduling (Resource Allocation)

1. Develop master schedule
   a. Estimate task duration
   b. Identify task precedence requirements
      i. Technical precedence
      ii. Resource-imposed precedence
      iii. Procedural precedence
   c. Use analytical models
      i. CPM
      ii. PERT
      iii. Gantt chart
      iv. Optimization models

Control (Tracking, Reporting, and Correction)

1. Establish guidelines for tracking, reporting, and control
   a. Define data requirements
      i. Data categories
      ii. Data characterization
      iii. Measurement scales
   b. Develop data documentation
      i. Data update requirements
      ii. Data quality control
      iii. Establish data security measures
2. Categorize control points
   a. Schedule audit
      i. Activity network and Gantt charts
      ii. Milestones
      iii. Delivery schedule
   b. Performance audit
      i. Employee performance
      ii. Product quality
   c. Cost audit
      i. Cost containment measures
      ii. Percent completion versus budget depletion
3. Identify implementation process
   a. Comparison with targeted schedules
   b. Corrective course of action
      i. Rescheduling
      ii. Reallocation of resources

Termination (Close, Phaseout)

1. Conduct performance review
2. Develop strategy for follow-up projects
3. Arrange for personnel retention, release, and reassignment

Documentation

1. Document project outcome
2. Submit final report
3. Archive report for future reference

Value of Lean Implementation

Facing a Lean period in project management creates value in terms of figuring out how to eliminate or reduce operational waste that is inherent in many human-governed processes. It is a natural fact that having to make do with limited resources creates opportunities for resourcefulness and innovation, which requires an integrated systems view of what is available and what can be leveraged. The Lean principles that are now being embraced by business, industry, and government have been around for a long time. It is just that we are now being forced to implement Lean practices due to the escalating shortage of resources. It is unrealistic to expect that problems that have enrooted themselves in different parts of an organization can be solved by a single-point attack. Rather, a systematic probing of all the nooks and corners of the problem must be assessed and tackled in an integrated manner. Contrary to the contention in some technocratic circles that budget cuts will stifle innovation, it is a fact that a reduction of resources often forces more creativity in identifying wastes and leveraging opportunities that lie fallow in nooks and crannies of an organization. This is not an issue of wanting more for less. Rather, it is an issue of doing more with less. It is through a systems viewpoint that new opportunities to innovate can be spotted. Necessity does, indeed, spur invention.

PROJECT DESIGN DECISION ANALYSIS

Systems decision analysis facilitates a proper consideration of the essential elements of decisions in a project systems environment. These essential elements include the problem statement, information, performance measure, decision model, and an implementation of the decision. The recommended steps are enumerated as follows:

Step 1. Problem Statement

A problem involves choosing between competing, and probably conflicting, alternatives. The components of problem-solving in project management include the following:

Describing the problem (goals, performance measures)
Defining a model to represent the problem
Solving the model
Testing the solution
Implementing and maintaining the solution.

Problem definition is very crucial. In many cases, *symptoms* of a problem are more readily recognized than its *cause* and *location*. Even after the problem is accurately identified and defined, a benefit/cost analysis may be needed to determine if the cost of solving the problem is justified.

Step 2. Data and Information Requirements

Information is the driving force for the project decision process. Information clarifies the relative states of past, present, and future events. The collection, storage, retrieval, organization, and processing of raw date are important components for generating information. Without data, there can be no information. Without good information, there cannot be a valid decision. The essential requirements for generating information are as follows:

Ensuring that an effective data collection procedure is followed
Determining the type and the appropriate amount of data to collect
Evaluating the data collected with respect to information potential
Evaluating the cost of collecting the required data.

For example, suppose a manager is presented with a recorded fact that says, "Sales for the last quarter are 10,000 units." This constitutes ordinary data. There are many ways of using the aforementioned data to make a decision, depending on the manager's value system. An analyst, however, can ensure the proper use of the data by transforming it into information, such as "Sales of

10,000 units for the last quarter are within x percent of the targeted value." This type of information is more useful to the manager for decision-making.

Step 3. Performance Measure

A performance measure for the competing alternatives should be specified. The decision-maker assigns a perceived worth or value to the available alternatives. Setting measures of performance is crucial to the process of defining and selecting alternatives. Some performance measures commonly used in project management are project cost, completion time, resource usage, and stability in the workforce.

Step 4. Decision Model

A decision model provides the basis for the analysis and synthesis of information, and is the mechanism by which competing alternatives are compared. To be effective, a decision model must be based on a systematic and logical framework for guiding project decisions. A decision model can be a verbal, graphical, or mathematical representation of the ideas in the decision-making process. A project decision model should have the following characteristics:

Simplified representation of the actual situation
Explanation and prediction of the actual situation
Validity and appropriateness
Applicability to similar problems.

The formulation of a decision model involves three essential components:

Abstraction: Determining the relevant factors
Construction: Combining the factors into a logical model
Validation: Assuring that the model adequately represents the problem.

The basic types of decision models for project management are described next:

Descriptive models: These models are directed at describing a decision scenario and identifying the associated problem. For example, a project analyst might use a critical path method (CPM) network model to identify bottleneck tasks in a project.

Prescriptive models: These models furnish procedural guidelines for implementing actions. The triple C approach (Badiru, 2019), for example, is a model that prescribes the procedures for achieving communication, cooperation, and coordination in a project environment.

Predictive models: These models are used to predict future events in a problem environment. They are typically based on historical data about the problem situation. For example, a regression model based on past data may be used to predict future productivity gains associated with expected levels of resource allocation. Simulation models can be used when uncertainties exist in the task durations or resource requirements.

Satisficing models: These are models that provide trade-off strategies for achieving a satisfactory solution to a problem, within given constraints. Goal programming and other multicriteria techniques provide good satisficing solutions. For example, these models are helpful in cases where time limitations, resource shortages, and performance requirements constrain the implementation of a project.

Optimization models: These models are designed to find the best available solution to a problem subject to a certain set of constraints. For example, a linear programming model can be used to determine the optimal product mix in a production environment.

In many situations, two or more of the aforementioned models may be involved in the solution of a problem. For example, a descriptive model might provide insights into the nature of the problem; an optimization model might provide the optimal set of actions to take in solving the problem; a satisficing model might temper the optimal solution with reality; a prescriptive model might suggest the procedures for implementing the selected solution; and a predictive model might forecast a future outcome of the problem scenario.

Step 5. Making the Decision

Using the available data, information, and the decision model, the decision-maker will determine the real-world actions that are needed to solve the stated problem. A sensitivity analysis may be useful for determining what changes in parameter values might cause a change in the decision.

Step 6. Implementing the Decision

A decision represents the selection of an alternative that satisfies the objective stated in the problem statement. A good decision is useless until it is implemented. An important aspect of a decision is to specify how it is to be implemented. Selling the decision and the project to management requires a well-organized persuasive presentation. The way a decision is presented can directly influence whether or not it is adopted. The presentation of a decision should include at least the following: an executive summary, technical aspects of the decision, managerial aspects of the decision, resources required to implement the decision, cost of the decision, the timeframe for implementing the decision, and the risks associated with the decision.

PROJECT GROUP DECISION-MAKING

Systems decisions are often complex, diffuse, distributed, and poorly understood. No one person has all the information to make all decisions accurately. As a result, crucial decisions are made by a group of people. Some organizations use outside consultants with appropriate expertise to make recommendations for important decisions. Other organizations set up their own internal consulting groups without having to go outside the organization. Decisions can be made through linear responsibility, in which case one person makes the final decision based on inputs from other people. Decisions can also be made through shared responsibility, in which case, a group of people share the responsibility for making joint decisions. The major advantages of group decision-making are listed as follows:

1. Facilitation of a systems view of the problem environment.
2. Ability to share experience, knowledge, and resources. Many heads are better than one. A group will possess greater collective ability to solve a given decision problem.
3. Increased credibility. Decisions made by a group of people often carry more weight in an organization.
4. Improved morale. Personnel morale can be positively influenced because many people have the opportunity to participate in the decision-making process.

5. Better rationalization. The opportunity to observe other people's views can lead to an improvement in an individual's reasoning process.
6. Ability to accumulate more knowledge and facts from diverse sources.
7. Access to broader perspectives spanning different problem scenarios.
8. Ability to generate and consider alternatives from different perspectives.
9. Possibility for a broader-base involvement, leading to a higher likelihood of support.
10. Possibility for group leverage for networking, communication, and political clout.

In spite of the much-desired advantages, group decision-making does possess the risk of flaws. Some possible disadvantages of group decision-making are listed as follows:

1. Difficulty in arriving at a decision.
2. Slow operating timeframe.
3. Possibility for individuals' conflicting views and objectives.
4. Reluctance of some individuals in implementing the decision.
5. Potential for power struggle and conflicts among the group.
6. Loss of productive employee time.
7. Too much compromise may lead to less than optimal group output.
8. Risk of one individual dominating the group.
9. Overreliance on group process may impede agility of management to make decision fast.
10. Risk of dragging feet due to repeated and iterative group meetings.

Brainstorming

Brainstorming is a way of generating many new ideas. In brainstorming, the decision group comes together to discuss alternate ways of solving a problem. The members of the brainstorming group may be from different departments, may have different backgrounds and training, and may not even know one another. The diversity of the participants helps create a stimulating environment for generating different ideas from different viewpoints. The technique encourages free outward expression of new ideas no matter how farfetched the ideas might appear. No criticism of any new idea is permitted during the brainstorming session. A major concern in brainstorming is that extroverts may take control of the discussions. For this reason, an experienced and respected individual should manage the brainstorming discussions.

The group leader establishes the procedure for proposing ideas, keeps the discussions in line with the group's mission, discourages disruptive statements, and encourages the participation of all members.

After the group runs out of ideas, open discussions are held to weed out the unsuitable ones. It is expected that even the rejected ideas may stimulate the generation of other ideas, which may eventually lead to other favored ideas. Guidelines for improving brainstorming sessions are presented as follows:

Focus on a specific decision problem.
Keep ideas relevant to the intended decision.
Be receptive to all new ideas.
Evaluate the ideas on a relative basis after exhausting new ideas.
Maintain an atmosphere conducive to cooperative discussions.
Maintain a record of the ideas generated.

Delphi Method

The traditional approach to group decision-making is to obtain the opinion of experienced participants through open discussions. An attempt is made to reach a consensus among the participants. However, open group discussions are often biased because of the influence of subtle intimidation from dominant individuals. Even when the threat of a dominant individual is not present, opinions may still be swayed by group pressure. This is called the "bandwagon effect" of group decision-making.

The Delphi method attempts to overcome these difficulties by requiring individuals to present their opinions anonymously through an intermediary. The method differs from the other interactive group methods because it eliminates face-to-face confrontations. It was originally developed for forecasting applications, but it has been modified in various ways for application to different types of decision-making. The method can be quite useful for project management decisions. It is particularly effective when decisions must be based on a broad set of factors. The Delphi method is normally implemented as follows:

1. *Problem definition*: A decision problem that is considered significant is identified and clearly described.
2. *Group selection*: An appropriate group of experts or experienced individuals is formed to address the particular decision problem. Both internal and external experts may be involved in the Delphi process. A leading individual is appointed to serve as the

administrator of the decision process. The group may operate through the mail or gather together in a room. In either case, all opinions are expressed anonymously on paper. If the group meets in the same room, care should be taken to provide enough room so that each member does not have the feeling that someone may accidentally or deliberately observe their responses.

3. *Initial opinion poll*: The technique is initiated by describing the problem to be addressed in unambiguous terms. The group members are requested to submit a list of major areas of concern in their specialty areas as they relate to the decision problem.

4. *Questionnaire design and distribution*: Questionnaires are prepared to address the areas of concern related to the decision problem. The written responses to the questionnaires are collected and organized by the administrator. The administrator aggregates the responses in a statistical format. For example, the average, mode, and median of the responses may be computed. This analysis is distributed to the decision group. Each member can then see how his or her responses compare with the anonymous views of the other members.

5. *Iterative balloting*: Additional questionnaires based on the previous responses are passed to the members. The members submit their responses again. They may choose to alter or not to alter their previous responses.

6. *Silent discussions and consensus*: The iterative balloting may involve anonymous written discussions of why some responses are correct or incorrect. The process is continued until a consensus is reached. A consensus may be declared after five or six iterations of the balloting or when a specified percentage (e.g., 80%) of the group agrees on the questionnaires. If a consensus cannot be declared on a particular point, it may be displayed to the whole group with a note that it does not represent a consensus.

In addition to its use in technological forecasting, the Delphi method has been widely used in other general decision-making. Its major characteristics of anonymity of responses, statistical summary of responses, and controlled procedure make it a reliable mechanism for obtaining numeric data from subjective opinion. The major limitations of the Delphi method are as follows:

1. Its effectiveness may be limited in cultures where strict hierarchy, seniority, and age influence decision-making processes.

2. Some experts may not readily accept the contribution of nonexperts to the group decision-making process.

3. Since opinions are expressed anonymously, some members may take the liberty of making ludicrous statements. However, if the group composition is carefully reviewed, this problem may be avoided.

Nominal Group Technique

The nominal group technique is a silent version of brainstorming. It is a method of reaching consensus. Rather than asking people to state their ideas aloud, the team leader asks each member to jot down a minimum number of ideas, for example, five or six. A single list of ideas is then written on a chalkboard for the whole group to see. The group then discusses the ideas and weeds out some iteratively until a final decision is made. The nominal group technique is easier to control. Unlike brainstorming where members may get into shouting matches, the nominal group technique permits members to silently present their views. In addition, it allows introversive members to contribute to the decision without the pressure of having to speak out too often.

In all of the group decision-making techniques, an important aspect that can enhance and expedite the decision-making process is to require that members review all pertinent data before coming to the group meeting. This will ensure that the decision process is not impeded by trivial preliminary discussions. Some disadvantages of group decision-making are as follows:

1. Peer pressure in a group situation may influence a member's opinion or discussions.
2. In a large group, some members may not get to participate effectively in the discussions.
3. A member's relative reputation in the group may influence how well his or her opinion is rated.
4. A member with a dominant personality may overwhelm the other members in the discussions.
5. The limited time available to the group may create a time pressure that forces some members to present their opinions without fully evaluating the ramifications of the available data.
6. It is often difficult to get all members of a decision group together at the same time.

Despite the noted disadvantages, group decision-making definitely has many advantages that may nullify the shortcomings. The advantages as presented earlier will have varying levels of effect from one organization to another. The triple C principle presented in Chapter 2 may also be used to improve the success of decision teams. Team work can be enhanced in group decision-making by adhering to the following guidelines:

1. Get a willing group of people together.
2. Set an achievable goal for the group.
3. Determine the limitations of the group.
4. Develop a set of guiding rules for the group.
5. Create an atmosphere conducive to group synergism.
6. Identify the questions to be addressed in advance.
7. Plan to address only one topic per meeting.

For major decisions and long-term group activities, arrange for team training that allows the group to learn the decision rules and responsibilities together. The steps for the nominal group technique are as follows:

1. Silently generate ideas, in writing.
2. Record ideas without discussion.
3. Conduct group discussion for clarification of meaning, not argument.
4. Vote to establish the priority or rank of each item.
5. Discuss vote.
6. Cast final vote.

Interviews, Surveys, and Questionnaires

Interviews, surveys, and questionnaires are important information gathering techniques. They also foster cooperative working relationships. They encourage direct participation and inputs into project decision-making processes. They provide an opportunity for employees at the lower levels of an organization to contribute ideas and inputs for decision-making. The greater the number of people involved in the interviews, surveys, and questionnaires, the more valid the final decision. The following guidelines are useful for conducting interviews, surveys, and questionnaires to collect data and information for project decisions:

1. Collect and organize background information and supporting documents on the items to be covered by the interview, survey, or questionnaire.
2. Outline the items to be covered, and list the major questions to be asked.
3. Use a suitable medium of interaction and communication: telephone, fax, electronic mail, face-to-face, observation, meeting venue, poster, or memo.

4. Tell the respondent the purpose of the interview, survey, or questionnaire, and indicate how long it will take.
5. Use open-ended questions that stimulate ideas from the respondents.
6. Minimize the use of yes or no type of questions.
7. Encourage expressive statements that indicate the respondent's views.
8. Use the who, what, where, when, why, and how approach to elicit specific information.
9. Thank the respondents for their participation.
10. Let the respondents know the outcome of the exercise.

Multivote

Multivoting is a series of votes used to arrive at a group decision. It can be used to assign priorities to a list of items. It can be used at team meetings after a brainstorming session has generated a long list of items. Multivoting helps reduce such long lists to a few items, usually three to five. The steps for multivoting are as follows:

1. Take a first vote. Each person votes as many times as desired, but only once per item.
2. Circle the items receiving a relatively higher number of votes (i.e., majority vote) than the other items.
3. Take a second vote. Each person votes for a number of items equal to one-half the total number of items circled in step 2. Only one vote per item is permitted.
4. Repeat steps 2 and 3 until the list is reduced to three to five items depending on the needs of the group. It is not recommended to multivote down to only one item.
5. Perform further analysis of the items selected in step 4, if needed.

Design via Planning

The key to a successful project is good planning. Project planning provides the basis for the initiation, implementation, and termination of a project. It sets guidelines for specific project objectives, project structure, tasks, milestones, personnel, cost, equipment, performance, and problem resolutions. An analysis of what is needed and what is available should be conducted in the planning phase of new projects. The availability of technical expertise within the organization and outside the organization should be reviewed.

If subcontracting is needed, the nature of the contract should undergo a thorough analysis. The question of whether or not the project is needed at all should be addressed. The "make," "buy," "lease," "subcontract," or "do-nothing" alternatives should be compared as part of the project planning process. Here are some guidelines for systems-wide project plans:

1. View a project plan as having tentacles that stretch across the organization.
2. Use project plans to coordinate across functional boundaries.
3. Establish plans as the platform over which project control will be done later on.
4. Leverage the diverse personalities and skills within the project environment.
5. Make room for contingent re-planning due to scope changes.
6. Empower workers to manage at the activity level.
7. Identify value-creating tasks and complementing activities.
8. Define specific milestones to facilitate project tracking.
9. Use checklists, tables, charts, and other visual tools project to communicate the plan.
10. Establish project performance metrics.

Although planning is a specific starting step in the project life cycle, it actually stretches over all the other steps of project management. Planning and re-planning permeate the project management life cycle. The major knowledge areas of project management, as presented by the PMI, are administered in a structured outline covering six basic clusters. The implementation clusters represent five process groups that are followed throughout the project life cycle. Each cluster itself consists of several functions and operational steps. When the clusters are overlaid on the nine knowledge areas in the Project Management Book of Knowledge (PMBOK®), we obtain a two-dimensional matrix that spans 44 major process steps. In general, project life cycle defines the following:

1. Resources that will be needed in each phase of the project life cycle
2. Specific work to be accomplished in each phase of the project life cycle.

The major phases of project life cycle going from the conceptual phase through the close-out phase are summarized below:

Project Initiation Phase → Project Development Phase →

Project Implementation Phase → Project Closure Phase

It should be noted that project life cycle is distinguished from product life cycle. Project life cycle does not explicitly address operational issues, whereas product life cycle is mostly about operational issues starting from the product's delivery to the end of its useful life. Note that for technical projects, the shape of the life cycle curve may be expedited due to the rapid developments that often occur in technology-based activities. For example, for a high-technology project, the entire life cycle may be shortened, with a very rapid initial phase, even though the conceptualization stage may be very long.

Typical characteristics of project life cycle include the following:

1. Cost and staffing requirements are lowest at the beginning of the project and ramp up during the initial and development stages.
2. The probability of successfully completing the project is lowest at the beginning and highest at the end. This is because many unknowns (risks and uncertainties) exist at the beginning of the project. As the project nears its end, there are fewer opportunities for risks and uncertainties.
3. The risks to the project organization (project owner) are lowest at the beginning and highest at the end. This is because not much investment has gone into the project at the beginning, whereas much has been committed by the end of the project. There is a higher sunk cost manifested at the end of the project.
4. The ability of the stakeholders to influence the final project outcome (cost, quality, and schedule) is highest at the beginning and gets progressively lower toward the end of the project. This is intuitive because influence is best exerted at the beginning of an endeavor.
5. Value of scope changes decreases over time during the project life cycle while the cost of scope changes increases over time. The suggestion is to decide and finalize scope as early as possible. If there are to be scope changes, do them as early as possible.

The specific application context will determine the essential elements contained in the life cycle of the endeavor. Life cycles of business entities, products, and projects have their own nuances that must be understood and managed within the prevailing organizational strategic plan. The components of corporate, product, and project life cycles are summarized as follows:

Corporate (Business) Life Cycle

Policy planning → Needs identification → Business conceptualization
→ Realization → Portfolio management

Product Life Cycle

Feasibility studies → Development → Operations → Product obsolescence

Project Life Cycle

Initiation → Planning → Execution → Monitoring and control → Closeout

There is no strict sequence for the application of the knowledge areas to a specific project. The areas represent a mixed collection of processes that must be followed in order to achieve a successful project. Thus, some aspects of planning may be found under the knowledge area for communications. In a similar vein, a project may start with the risk management process before proceeding into the integration process. The knowledge areas provide general guidelines. Each project must adapt and tailor the recommended techniques to the specific need and unique circumstances of the project. PMI's PMBOK seeks to standardize project management terms and definitions by presenting a common lexicon for project management activities.

Specific strategic, operational, and tactical goals and objectives are embedded within each step in the loop. For example, "initiating" may consist of project conceptualization and description. Part of "executing" may include resource allocation and scheduling. "Monitoring" may involve project tracking, data collection, and parameter measurement. "Controlling" implies taking corrective action based on the items that are monitored and evaluated. "Closing" involves phasing out or terminating a project. Closing does not necessarily mean a death sentence for a project as the end of one project may be used as the stepping stone to initiate the next series of endeavors.

In the initial stage of project planning, the internal and external factors that influence the project should be determined and given priority weights. Examples of internal influences on project plans include the following:

Infrastructure
Project scope
Labor relations
Project location
Project leadership
Organizational goal
Management approach
Technical personnel supply
Resource and capital availability.

In addition to internal factors, a project plan can be influenced by external factors. An external factor may be the sole instigator of a project, or it may manifest itself in combination with other external and internal factors. Such external factors include the following:

Public needs
Market needs
National goals
Industry stability
State of technology
Industrial competition
Government regulations.

TIME-COST-PERFORMANCE TRADE-OFFS IN PROJECT DESIGN

Project goals determine the nature of project planning. Project goals may be specified in terms of time (schedule), cost (resources), or performance (output). A project can be simple or complex. While simple projects may not require the whole array of project management tools, complex projects may not be successful without all the tools. Project management techniques are applicable to a wide collection of problems ranging from manufacturing to medical services.

The techniques of project management can help achieve goals relating to better product quality, improved resource utilization, better customer relations, higher productivity, and fulfillment of due dates. These can be expressed in terms of the following project constraints:

Performance specifications
Schedule requirements
Cost limitations.

Project planning determines the nature of actions and responsibilities needed to achieve the project goal. It entails the development of alternate courses of action and the selection of the best action to achieve the objectives making up the goal. Planning determines what needs to be done, by whom, and when. Whether it is done for long-range (strategic) purposes or for short-range (operational) purposes, planning should be one of the first steps of project management.

STRATEGIC LEVELS OF PLANNING

Decisions involving strategic planning lay the foundation for the successful implementation of projects. Planning forms the basis for all actions. Strategic decisions may be divided into three strategy levels: *supralevel planning*, *macrolevel planning*, and *microlevel planning*.

Supralevel planning: Planning at the supralevel deals with the big picture of how the project fits the overall and long-range organizational goals. Questions faced at this level concern potential contributions of the project to the welfare of the organization, its effect on the depletion of company resources, required interfaces with other projects within and outside the organization, risk exposure, management support for the project, concurrent projects, company culture, market share, shareholder expectations, and financial stability.

Macrolevel planning: Planning decisions at the macrolevel address the overall planning within the project boundary. The scope of the project and its operational interfaces should be addressed at this level. Questions faced at the macrolevel include goal definition, project scope, availability of qualified personnel, resource availability, project policies, communication interfaces, budget requirements, goal interactions, deadline, and conflict resolution strategies.

Microlevel planning: The microlevel deals with detailed operational plans at the task levels of the project. Definite and explicit tactics for accomplishing specific project objectives are developed at the microlevel. The concept of management by objective (MBO) may be particularly effective at this level. MBO permits each project member to plan his or her own work at the microlevel. Factors to be considered at the microlevel of project decisions include scheduled time, training requirements, required tools, task procedures, reporting requirements, and quality requirements.

Project decisions at the three levels defined previously will involve numerous personnel within the organization with various types and levels of expertise. In addition to the conventional roles of the project manager, specialized roles may be developed within the project scope. Such roles include the following:

1. *Technical specialist:* This person will have responsibility for addressing specific technical requirements of the project. In a large project, there will typically be several technical specialists working together to solve project problems.
2. *Operations integrator:* This person will be responsible for making sure that all operational components of the project interface

correctly to satisfy project goals. This person should have good technical awareness and excellent interpersonal skills.

3. *Project specialist:* This person has specific expertise related to the specific goals and requirements of the project. Even though a technical specialist may also serve as a project specialist, the two roles should be distinguished. A general electrical engineer may be a technical specialist on the electronic design components of a project. However, if the specific setting of the electronics project is in the medical field, then an electrical engineer with expertise in medical operations may be needed to serve as the project specialist.

COMPONENTS OF A PROJECT DESIGN PLAN

Planning is an ongoing process that is conducted throughout the project life cycle. Initial planning may relate to overall organizational efforts. This is where specific projects to be undertaken are determined. Subsequent planning may relate to specific objectives of the selected project. In general, a project plan should consist of the following components:

1. *Summary of project plan:* This is a brief description of what is planned. Project scope and objectives should be enumerated. The critical constraints on the project should be outlined. The types of resources required and available should be specified. The summary should include a statement of how the project complements organizational and national goals, budget size, and milestones.

2. *Objectives:* The objectives should be very detailed in outlining what the project is expected to achieve and how the expected achievements will contribute to the overall goals of a project. The performance measures for evaluating the achievement of the objectives should be specified.

3. *Approach:* The managerial and technical methodologies of implementing the project should be specified. The managerial approach may relate to project organization, communication network, approval hierarchy, responsibility, and accountability. The technical approach may relate to company experience on previous projects and currently available technology.

4. *Policies and procedures:* Development of a project policy involves specifying the general guidelines for carrying out tasks within the project. Project procedure involves specifying the detailed method for implementing a given policy relative to the tasks needed to achieve the project goal.

5. *Contractual requirements:* This portion of the project plan should outline reporting requirements, communication links, customer specifications, performance specifications, deadlines, review process, project deliverables, delivery schedules, internal and external contacts, data security, policies, and procedures. This section should be as detailed as practically possible. Any item that has the slightest potential of creating problems later should be documented.

6. *Project schedule:* The project schedule signifies the commitment of resources against time in pursuit of project objectives. A project schedule should specify when the project will be initiated and when it is expected to be completed. The major phases of the project should be identified. The schedule should include reliable time estimates for project tasks. The estimates may come from knowledgeable personnel, past records, or forecasting. Task milestones should be generated on the basis of objective analysis rather than arbitrary stipulations. The schedule in this planning stage constitutes the master project schedule. Detailed activity schedules should be generated under specific project functions.

7. *Resource requirements:* Project resources, budget, and costs are to be documented in this section of the project plan. Capital requirements should be specified by tasks. Resources may include personnel, equipment, and information. Special personnel skills, hiring, and training should be explained. Personnel requirements should be aligned with schedule requirements so as to ensure their availability when needed. Budget size and source should be presented. The basis for estimating budget requirements should be justified, and the cost allocation and monitoring approach should be shown.

8. *Performance measures:* Measures of evaluating project progress should be developed. The measures may be based on standard practices or customized needs. The method of monitoring, collecting, and analyzing the measures should also be specified. Corrective actions for specific undesirable events should be outlined.

9. *Contingency plans:* Courses of actions to be taken in the case of undesirable events should be predetermined. Many projects have failed simply because no plans have been developed for emergency situations. In the excitement of getting a project under way, it is often easy to overlook the need for contingency plans.

10. *Tracking, reporting, and auditing:* These involve keeping track of the project plans, evaluating tasks, and scrutinizing the records of the project.

Planning for large projects may include a statement about the feasibility of subcontracting part of the project work. Subcontracting may be needed for various reasons including lower cost, higher efficiency, and logistical convenience.

SYSTEMS HIERARCHY FOR PROJECT DESIGN

The traditional concepts of systems analysis are applicable to the project process. The definitions of a project system and its components are presented next:

System: A project system consists of interrelated elements organized for the purpose of achieving a common goal. The elements are organized to work synergistically to generate a unified output that is greater than the sum of the individual outputs of the components.

Program: A program is a very large and prolonged undertaking. Such endeavors often span several years. Programs are usually associated with particular systems. For example, we may have a space exploration program within a national defense system.

Project: A project is a time-phased effort of much smaller scope and duration than a program. Programs are sometimes viewed as consisting of a set of projects. Government projects are often called *programs* because of their broad and comprehensive nature. Industry tends to use the term *project* because of the short-term and focused nature of most industrial efforts.

Task: A task is a functional element of a project. A project is composed of a sequence of tasks that all contribute to the overall project goal.

Activity: An activity can be defined as a single element of a project. Activities are generally smaller in scope than tasks. In a detailed analysis of a project, an activity may be viewed as the smallest, practically indivisible work element of the project. For example, we can regard a manufacturing plant as a system. A plantwide endeavor to improve productivity can be viewed as a program. The installation of a flexible manufacturing system is a project

within the productivity improvement program. The process of identifying and selecting equipment vendors is a task, and the actual
process of placing an order with a preferred vendor is an activity.

The emergence of systems development has had an extensive effect on project management in recent years. A system can be defined as a collection of
interrelated elements brought together to achieve a specified objective. In a
management context, the purposes of a system are to develop and manage
operational procedures and to facilitate an effective decision-making process.
Some of the common characteristics of a system include the following:

1. Interaction with the environment
2. Objective
3. Self-regulation
4. Self-adjustment.

Representative components of a project system are the organizational subsystem, planning subsystem, scheduling subsystem, information management
subsystem, control subsystem, and project delivery subsystem. The primary
responsibilities of project analysts involve ensuring the proper flow of information throughout the project system. The classical approach to the decision
process follows rigid lines of organizational charts. By contrast, the systems
approach considers all the interactions necessary among the various elements
of an organization in the decision process.

The various elements (or subsystems) of the organization act simultaneously in a separate but interrelated fashion to achieve a common goal. This
synergism helps to expedite the decision process and to enhance the effectiveness of decisions. The supporting commitments from other subsystems of
the organization serve to counterbalance the weaknesses of a given subsystem. Thus, the overall effectiveness of the system is greater than the sum of
the individual results from the subsystems.

The increasing complexity of organizations and projects makes the systems approach essential in today's management environment. As the number
of complex projects increases, there will be an increasing need for project
management professionals who can function as systems integrators. Project
management techniques can be applied to the various stages of implementing
a system as shown in the following guidelines:

1. *Systems definition*: Define the system and associated problems using
   keywords that signify the importance of the problem to the overall
   organization. Locate experts in this area who are willing to contribute to the effort. Prepare and announce the development plan.

2. *Personnel assignment*: The project group and the respective tasks should be announced, a qualified project manager should be appointed, and a solid line of command should be established and enforced.

3. *Project initiation*: Arrange an organizational meeting during which a general approach to the problem should be discussed. Prepare a specific development plan, and arrange for the installation of needed hardware and tools.

4. *System prototype*: Develop a prototype system, test it, and learn more about the problem from the test results.

5. *Full system development*: Expand the prototype to a full system, evaluate the user interface structure, and incorporate user training facilities and documentation.

6. *System verification*: Get experts and potential users involved, ensure that the system performs as designed, and debug the system as needed.

7. *System validation*: Ensure that the system yields expected outputs. Validate the system by evaluating performance level, such as percentage of success in so many trials, measuring the level of deviation from expected outputs, and measuring the effectiveness of the system output in solving the problem.

8. *System integration*: Implement the full system as planned, ensure the system can coexist with systems already in operation, and arrange for technology transfer to other projects.

9. *System maintenance*: Arrange for continuing maintenance of the system. Update solution procedures as new pieces of information become available. Retain responsibility for system performance or delegate to well-trained and authorized personnel.

10. *Documentation*: Prepare full documentation of the system, prepare a user's guide, and appoint a user consultant.

Systems integration permits sharing of resources. Physical equipment, concepts, information, and skills may be shared as resources. Systems integration is now a major concern of many organizations. Even some of the organizations that traditionally compete and typically shun cooperative efforts are beginning to appreciate the value of integrating their operations. For these reasons, systems integration has emerged as a major interest in business. Systems integration may involve the physical integration of technical components, objective integration of operations, conceptual integration of management processes, or a combination of any of these.

Systems integration involves the linking of components to form subsystems and the linking of subsystems to form composite systems within a single department and/or across departments. It facilitates the coordination of technical and managerial efforts to enhance organizational functions, reduce cost, save energy, improve productivity, and increase the utilization of resources. Systems integration emphasizes the identification and coordination of the interface requirements among the components in an integrated system. The components and subsystems operate synergistically to optimize the performance of the total system. Systems integration ensures that all performance goals are satisfied with a minimum expenditure of time and resources. Integration can be achieved in several forms including the following:

1. *Dual-use integration*: This involves the use of a single component by separate subsystems to reduce both the initial cost and the operating cost during the project life cycle.
2. *Dynamic resource integration*: This involves integrating the resource flows of two normally separate subsystems so that the resource flow from one to or through the other minimizes the total resource requirements in a project.
3. *Restructuring of functions*: This involves the restructuring of functions and reintegration of subsystems to optimize costs when a new subsystem is introduced into the project environment.

Systems integration is particularly important when introducing new technology into an existing system. It involves coordinating new operations to coexist with existing operations. It may require the adjustment of functions to permit the sharing of resources, development of new policies to accommodate product integration, or realignment of managerial responsibilities. It can affect both hardware and software components of an organization. Presented in the following list are guidelines and important questions relevant for systems integration:

What are the unique characteristics of each component in the integrated system?
How do the characteristics complement one another?
What physical interfaces exist among the components?
What data/information interfaces exist among the components?
What ideological differences exist among the components?
What are the data flow requirements for the components?
Are there similar integrated systems operating elsewhere?
What are the reporting requirements in the integrated system?
Are there any hierarchical restrictions on the operations of the components of the integrated system?
What internal and external factors are expected to influence the integrated system?

How can the performance of the integrated system be measured?

What benefit/cost documentations are required for the integrated system?

What is the cost of designing and implementing the integrated system?

What are the relative priorities assigned to each component of the integrated system?

What are the strengths of the integrated system?

What are the weaknesses of the integrated system?

What resources are needed to keep the integrated system operating satisfactorily?

Which section of the organization will have primary responsibility for the operation of the integrated system?

What are the quality specifications and requirements for the integrated systems?

The integrated approach to project management recommended in this book is represented by the flowchart in Figure 2.1.

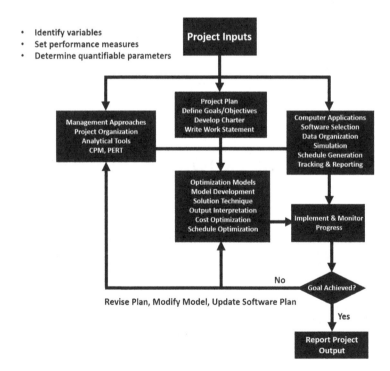

FIGURE 2.1   Project design and execution flowchart.

The process starts with a managerial analysis of the project effort. Goals and objectives are defined, a mission statement is written, and the statement of work is developed. After these, traditional project management approaches, such as the selection of an organization structure, are employed. Conventional analytical tools including the CPM and the program evaluation and review technique (PERT) are then mobilized. The use of optimization models is then called upon as appropriate. Some of the parameters to be optimized are cost, resource allocation, and schedule length. It should be understood that not all project parameters will be amenable to optimization. The use of commercial project management software should start only after the managerial functions have been completed. Some project management software have built-in capabilities for the planning and optimization needs.

A frequent mistake in project management is the rush to use a project management software without first completing the planning and analytical studies required by the project. Project management software should be used as a management tool, the same way a word processor is used as a writing tool. It will not be effective to start using the word processor without first organizing the thoughts about what is to be written. Project management is much more than just the project management software. If project management is carried out in accordance with the integration approach presented in the flowchart, the odds of success will be increased. Of course, the structure of the flowchart should not be rigid. Flows and interfaces among the blocks in the flowchart may need to be altered or modified depending on specific project needs.

WORK BREAKDOWN STRUCTURE

The concept of "divide and conquer" works in many problem-solving challenges. It also works in breaking down a project into its components parts that are easier to manage in smaller chunks.

The core of a project design is the work breakdown structure (WBS), which provides a layout for work and data interfaces on the project. WBS represents a family tree hierarchy of project operations required to accomplish project objectives. It is particularly useful for purposes of planning, scheduling, and control. Tasks that are contained in the WBS collectively describe the overall project. The tasks may involve physical products (e.g., steam generators), services (e.g., testing), and data (e.g., reports, sales data). The WBS serves to describe the link between the end objective and the operations required to reach that objective. It shows work elements in the conceptual framework for planning and controlling. The objective of developing a WBS is to study the elemental components of a project in detail. It permits the implementation of the "divide and conquer" concepts. Overall project planning and control can be improved by using a WBS approach. A large project may be broken down into smaller subprojects which may, in turn, be systematically broken down into task groups.

Individual components in a WBS are referred to as WBS elements, and the hierarchy of each is designated by a level identifier. Elements at the same level of subdivision are said to be of the same WBS level. Descending levels provide increasingly detailed definition of project tasks. The complexity of a project and the degree of control desired determine the number of levels in the WBS. An example of a WBS is shown in Figure 2.2.

Each WBS component is successively broken down into smaller details at lower levels. The process may continue until specific project activities are reached. The basic approach for preparing a WBS is as follows:

Level 1: It contains only the final project purpose. This item should be identifiable directly as an organizational budget item.

Level 2: It contains the major subsections of the project. These subsections are usually identified by their contiguous location or by their related purposes.

Level 3: It contains definable components of the level 2 subsections.

Subsequent levels are constructed in more specific detail depending on the level of control desired. If a complete WBS becomes too crowded, separate

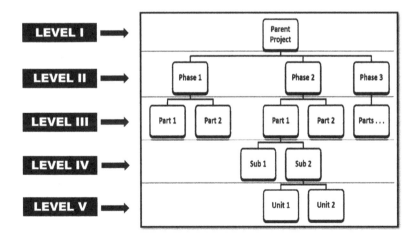

FIGURE 2.2 Work breakdown structure.

WBSs may be drawn for the level 2 components. A *specification of work* or WBS summary should normally accompany the WBS. A statement of work is a narrative of the work to be done. It should include the objectives of the work, its nature, resource requirements, and tentative schedule. Each WBS element is assigned a code that is used for its identification throughout the project life cycle. Alphanumeric codes may be used to indicate element level as well as component group.

CONCLUSIONS

This chapter has presented concepts and guidance relevant for project design in the context of DEJI Systems Model. In this case, design is defined generically to cover project concept, technical structure, qualitative characteristics, and scope. With good designs comes more effective decision processes about project elements and activity management strategies.

REFERENCE

Badiru, A. B. (2019). *Project management: systems, principles, and applications* (2nd ed.). Boca Raton, FL: Taylor & Francis Group/CRC Press.

Project Evaluation

3

When project curiosity is established, the
urge to learn more develops.
– Adedeji Badiru

ELEMENTS OF PROJECT EVALUATION

Project evaluation is the second stage of the DEJI Systems Model. It focuses on developing a comprehensive profile of the project. Some elements of the evaluation are qualitative, while some are quantitative. The goal of this stage of the systems model is to have a better understanding of the project from the perspective of both internal and external stakeholders. A project that is not fully understood will run into execution problems later on in the implementation process.

USING TRIPLE C MODEL FOR PROJECT EVALUATION

The Triple C model (Badiru, 2008) is an effective project planning tool. The model states that project management can be enhanced by implementing it within the integrated functions of the following elements:

- Communication
- Cooperation
- Coordination.

The model facilitates a systematic approach to project planning, organizing, scheduling, and control. It could be an effective tool for project evaluation.

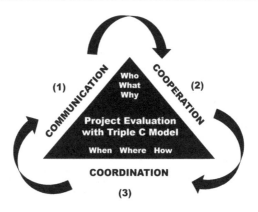

FIGURE 3.1 Illustration of the Triple C model.

The Triple C model can be implemented for project planning, scheduling, and control purposes. The model is shown graphically in Figure 3.1. Good communication leads to solid cooperation, which paves the way for a sustainable coordination and then facilitates additional communication.

The Triple C model highlights what must be done and when. It can also help to identify the resources (personnel, equipment, facilities, etc.) required for each effort. It points out important questions such as the following:

- Does each project participant know what the objective is?
- Does each project participant know his or her role in achieving the objective?
- What obstacles may prevent a participant from playing his or her role effectively?

COMMUNICATION

Communication makes working together possible. The communication function of project management involves making all those concerned become aware of project requirements and progress. Those who will be affected by the project directly or indirectly, as direct participants or as beneficiaries, should be informed as appropriate regarding the following:

Scope of the project
Personnel contribution required
Expected cost and merits of the project

Project organization and implementation plan
Potential adverse effects if the project should fail
Alternatives, if any, for achieving the project goal
Potential direct and indirect benefits of the project.

The communication channel must be kept open throughout the project life cycle. In addition to internal communication, appropriate external sources should also be consulted. The project manager must

Exude commitment to the project
Utilize the communication responsibility matrix
Facilitate multichannel communication interfaces
Identify internal and external communication needs
Resolve organizational and communication hierarchies
Encourage both formal and informal communication links.

In order for a responsibility matrix to be effective, it must be disseminated appropriately so that all who need to know are, indeed, aware of the project and their respective roles in order to remove ambiguity. When clear communication is maintained between management and employees and among peers, many project problems can be averted. Project communication may be carried out in one or more of the following formats:

One-to-many
One-to-one
Many-to-one
Written and formal
Written and informal
Oral and formal
Oral and informal
Nonverbal gesture.

Good communication is effected when what is implied is perceived as intended. Effective communications are vital to the success of any project. Despite the awareness that proper communications form the blueprint for project success, many organizations still fail in their communication functions. The study of communication is complex. Factors that influence the effectiveness of communication within a project organization structure include the following:

1. *Personal perception*: Each person perceives events on the basis of personal psychological, social, cultural, and experiential background. As a result, no two people can interpret a given event the same way.

The nature of events is not always the critical aspect of a problem situation. Rather, the problem is often the different perceptions of the different people involved.

2. *Psychological profile*: The psychological makeup of each person determines personal reactions to events or words. Thus, individual needs and level of thinking will dictate how a message is interpreted.

3. *Social environment*: Communication problems sometimes arise because people have been conditioned by their prevailing social environment to interpret certain things in unique ways. Vocabulary, idioms, organizational status, social stereotypes, and economic situation are among the social factors that can thwart effective communication.

4. *Cultural background*: Cultural differences are among the most pervasive barriers to project communications, especially in today's multinational organizations. Language and cultural idiosyncrasies often determine how communication is approached and interpreted.

5. *Semantic and syntactic factors*: Semantic and syntactic barriers to communications usually occur in written documents. Semantic factors are those that relate to the intrinsic knowledge of the subject of the communication. Syntactic factors are those that relate to the form in which the communication is presented. The problems created by these factors become acute in situations where response, feedback, or reaction to the communication cannot be observed.

6. *Organizational structure*: Frequently, the organization structure in which a project is conducted has a direct influence on the flow of information and, consequently, on the effectiveness of communication. Organization hierarchy may determine how different personnel levels perceive a given communication.

7. *Communication media*: The method of transmitting a message may also affect the value ascribed to the message and, consequently, how it is interpreted or used. The common barriers to project communications are listed as follows:

Inattentiveness
Lack of organization
Outstanding grudges
Preconceived notions
Ambiguous presentation
Emotions and sentiments
Lack of communication feedback

Sloppy and unprofessional presentation
Lack of confidence in the communicator
Lack of confidence by the communicator
Low credibility of communicator
Unnecessary technical jargon
Too many people involved
Untimely communication
Arrogance or imposition
Lack of focus.

Some suggestions on improving the effectiveness of communication are presented next. The recommendations may be implemented as appropriate for any of the forms of communication listed earlier. The recommendations are for both the communicator and the audience.

1. Never assume that the integrity of the information sent will be preserved, as the information passes through several communication channels. Information is generally filtered, condensed, or expanded by the receivers before relaying it to the next destination. When preparing a communication that needs to pass through several organization structures, one safeguard is to compose the original information in a concise form to minimize the need for re-composition.

2. Give the audience a central role in the discussion. A leading role can help make a person feel a part of the project effort and responsible for the project's success. He or she can then have a more constructive view of project communication.

3. Do homework and think through the intended accomplishment of the communication. This helps eliminate trivial and inconsequential communication efforts.

4. Carefully plan the organization of the ideas embodied in the communication. Use indexing or points of reference whenever possible. Grouping ideas into related chunks of information can be particularly effective. Present the short message first. Short messages help create focus, maintain interest, and prepare the mind for the longer messages to follow.

5. Highlight why the communication is of interest and how it is intended to be used. Full attention should be given to the content of the message with regard to the prevailing project situation.

6. Elicit the support of those around you by integrating their ideas into the communication. The more people feel they have contributed to the issue, the more expeditious they are in soliciting the

cooperation of others. The effect of the multiplicative rule can quickly garner support for the communication purpose.

7. Be responsive to the feelings of others. It takes two to communicate. Anticipate and appreciate the reactions of members of the audience. Recognize their operational circumstances and present your message in a form they can relate to.

8. Accept constructive criticism. Nobody is infallible. Use criticism as a springboard to higher communication performance.

9. Exhibit interest in the issue in order to arouse the interest of your audience. Avoid delivering your message as a matter of a routine organizational requirement.

10. Obtain and furnish feedback promptly. Clarify vague points with examples.

11. Communicate at the appropriate time, at the right place, to the right people.

12. Reinforce words with positive action. Never promise what cannot be delivered. Value your credibility.

13. Maintain eye contact in oral communication, and read the facial expressions of your audience to obtain real-time feedback.

14. Concentrate on listening as much as speaking. Evaluate both the implicit and explicit meanings of statements.

15. Document communication transactions for future references.

16. Avoid asking questions that can be answered yes or no. Use relevant questions to focus the attention of the audience. Use questions that make people reflect upon their words, such as, "How do you think this will work?" compared to "Do you think this will work?"

17. Avoid patronizing the audience. Respect their judgment and knowledge.

18. Speak and write in a controlled tempo. Avoid emotionally charged voice inflections.

19. Create an atmosphere for formal and informal exchanges of ideas.

20. Summarize the objectives of the communication and how they will be achieved.

A communication responsibility matrix shows the linking of sources of communication and targets of communication. Cells within the matrix indicate the subject of the desired communication. There should be at least one filled cell in each row and each column of the matrix. This assures that each individual of a department has at least one communication source or target associated with him or her. With a communication responsibility matrix, a clear understanding of what needs to be communicated to whom can be developed.

Communication in a project environment can take any of several forms. The specific needs of a project may dictate the most appropriate mode. Three popular computer communication modes are discussed next in the context of communicating data and information for project management.

Simplex communication: This is a unidirectional communication arrangement in which one project entity initiates communication to another entity or individual within the project environment. The entity addressed in the communication does not have a mechanism or capability for responding to the communication. An extreme example of this is a one-way, top-down communication from top management to the project personnel. In this case, the personnel have no communication access or input to top management. A budget-related example is a case where top management allocates budget to a project without requesting and reviewing the actual needs of the project. Simplex communication is common in authoritarian organizations.

Half-duplex communication: This is a bidirectional communication arrangement whereby one project entity can communicate with another entity and receive a response within a certain time lag. Both entities can communicate with each other but not at the same time. An example of half-duplex communication is a project organization that permits communication with top management without a direct meeting. Each communicator must wait for a response from the target of the communication. Request and allocation without a budget meeting is another example of half-duplex data communication in project management.

Full-duplex communication: This involves a communication arrangement that permits a dialogue between the communicating entities. Both individuals and entities can communicate with each other at the same time or face to face. As long as there is no clash of words, this appears to be the most receptive communication mode. It allows participative project planning in which each project personnel has an opportunity to contribute to the planning process.

Each member of a project team needs to recognize the nature of the prevailing communication mode in the project. Management must evaluate the prevailing communication structure and attempt to modify it if necessary to enhance project functions. An evaluation of who is to communicate with whom about what may help improve the project data/information communication process. A communication matrix may include notations about the desired modes of communication between individuals and groups in the project environment. The types of communication, cooperation, and coordination that can be in effect in a project environment are summarized as follows:

Types of communication
Verbal
Written

Body language
Visual tools (i.e., graphical tools)
One-to-one
One-to-many
Many-to-one

Types of cooperation
Proximity
Functional
Professional
Social
Power influence
Social
Authority influence
Hierarchical
Lateral
Cooperation by intimidation
Cooperation by enticement

Types of coordination
Teaming
Delegation
Supervision
Partnership
Token-passing
Baton hand-off

Complexity of Multi-Person Communication

Communication complexity increases with an increase in the number of communication channels. It is one thing to wish to communicate freely, but it is another thing to contend with the increased complexity when more people are involved. The statistical formula of combination can be used to estimate the complexity of communication as a function of the number of communication channels or number of participants. The combination formula is used to calculate the number of possible combinations of r objects from a set of n objects. This is written as

$$_nC_r = \frac{n!}{r![n-r]!}$$

In the case of communication, for illustration purposes, we assume communication is between two members of a team at a time. That is, combination of two from n team members. That is, number of possible combinations of two members out of a team of n people. Thus, the formula for communication complexity reduces to the expression as follows, after some of the computation factors cancel out:

$$_nC_2 = \frac{n(n-1)}{2}$$

In a similar vein, Badiru (2008) introduced a formula for cooperation complexity based on the statistical concept of permutation. Permutation is the number of possible arrangements of k objects taken from a set of n objects. The permutation formula is written as

$$_nP_k = \frac{n!}{(n-k)!}$$

Thus, the number of possible permutations of two members out of a team of n members is estimated as

$$_nP_2 = n(n-1)$$

Permutation formula is used for cooperation because cooperation is bidirectional. Full cooperation requires that if A cooperates with B, then B must cooperate with A. But, A cooperating with B does not necessarily imply B cooperating with A. In notational form, that is,

A $\rightarrow$ B does not necessarily imply B $\rightarrow$ A

Figure 3.2 shows the relative plots of communication complexity and cooperation complexity as a function of project team size, n.

PROJECT COOPERATION

The cooperation of the project personnel must be explicitly elicited. Merely voicing a consent for a project is not enough assurance of full cooperation. The participants and beneficiaries of the project must be convinced of the merits of the project. Some of the factors that influence cooperation in a project environment include personnel requirements, resource requirements,

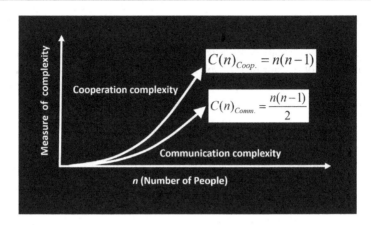

FIGURE 3.2 Communication and cooperation complexity as a function of team size.

budget limitations, past experiences, conflicting priorities, and lack of uniform organizational support. A structured approach to seeking cooperation should clarify the following:

Cooperative efforts required
Precedents for future projects
Implication of lack of cooperation
Criticality of cooperation to project success
Organizational impact of cooperation
Time frame involved in the project
Rewards of good cooperation.

Cooperation is a basic virtue of human interaction. More projects fail due to a lack of cooperation and commitment than any other project factors. To secure and retain the cooperation of project participants, you must elicit a positive first reaction to the project. The most positive aspects of a project should be the first items of project communication. For project management, there are different types of cooperation that should be understood.

Functional cooperation: This is cooperation induced by the nature of the functional relationship between two groups. The two groups may be required to perform related functions that can only be accomplished through mutual cooperation.

Social cooperation: This is the type of cooperation effected by the social relationship between two groups. The prevailing social relationship motivates cooperation that may be useful in getting project work done.

Legal cooperation: Legal cooperation is the type of cooperation that is imposed through some authoritative requirement. In this case, the participants may have no choice other than to cooperate.

Administrative cooperation: This is cooperation brought on by administrative requirements that make it imperative that two groups work together on a common goal.

Associative cooperation: This type of cooperation may also be referred to as collegiality. The level of cooperation is determined by the association that exists between two groups.

Proximity cooperation: Cooperation due to the fact that two groups are geographically close is referred to as proximity cooperation. Being close makes it imperative that the two groups work together.

Dependency cooperation: This is cooperation caused by the fact that one group depends on another group for some important aspect. Such dependency is usually of a mutual two-way nature. One group depends on the other for one thing, while the latter group depends on the former for some other thing.

Imposed cooperation: In this type of cooperation, external agents must be employed to induce cooperation between two groups. This is applicable for cases where the two groups have no natural reason to cooperate. This is where the approaches presented earlier for seeking cooperation can become very useful.

Lateral cooperation: Lateral cooperation involves cooperation with peers and immediate associates. Lateral cooperation is often easy to achieve because existing lateral relationships create a conducive environment for project cooperation.

Vertical cooperation: Vertical or hierarchical cooperation refers to cooperation that is implied by the hierarchical structure of the project. For example, subordinates are expected to cooperate with their vertical superiors.

Whichever type of cooperation is available in a project environment, the cooperative forces should be channeled toward achieving project goals. Documentation of the prevailing level of cooperation is useful for winning further support for a project. Clarification of project priorities will facilitate personnel cooperation. Relative priorities of multiple projects should be specified so that a project that is of high priority to one segment of an organization is also of high priority to all groups within the organization. Some guidelines for securing cooperation for most projects are as follows:

Establish achievable goals for the project.
Clearly outline the individual commitments required.
Integrate project priorities with existing priorities.

Eliminate the fear of job loss due to industrialization.
Anticipate and eliminate potential sources of conflict.
Use an open-door policy to address project grievances.
Remove skepticism by documenting the merits of the project.

Commitment: Cooperation must be supported with commitment. To cooperate is to support the ideas of a project. To commit is to willingly and actively participate in project efforts again and again through the thick and thin of the project. Provision of resources is one way that management can express commitment to a project.

Triple C + Commitment = Project success

PROJECT COORDINATION USING A RESPONSIBILITY MATRIX

After the communication and cooperation functions have successfully been initiated, the efforts of the project personnel must be coordinated. Coordination facilitates harmonious organization of project efforts. The construction of a responsibility chart can be very helpful at this stage. A responsibility chart is a matrix consisting of columns of individuals or functional departments and rows of required actions. Cells within the matrix are filled with relationship codes that indicate who is responsible for what.

The matrix helps avoid neglecting crucial communication requirements and obligations. It can help resolve questions such as

Who is to do what?
How long will it take?
Who is to inform whom of what?
Whose approval is needed for what?
Who is responsible for which results?
What personnel interfaces are required?
What support is needed from whom and when?

When implemented as an integrated process, the Triple C model can help avoid conflicts in a project. When conflicts do develop, it can help in resolving the conflicts. Several sources of conflicts can exist in large projects. Some of these are discussed next.

Schedule conflict: Conflicts can develop because of improper timing or sequencing of project tasks. This is particularly common in large multiple projects. Procrastination can lead to having too much to do at once, thereby creating a clash of project functions and discord among project team members. Inaccurate estimates of time requirements may lead to infeasible activity schedules. Project coordination can help avoid schedule conflicts.

Cost conflict: Project cost may not be generally acceptable to the clients of a project. This will lead to project conflict. Even if the initial cost of the project is acceptable, a lack of cost control during project implementation can lead to conflicts. Poor budget allocation approaches and the lack of a financial feasibility study will cause cost conflicts later on in a project. Communication and coordination can help prevent most of the adverse effects of cost conflicts.

Performance conflict: If clear performance requirements are not established, performance conflicts will develop. Lack of clearly defined performance standards can lead each person to evaluate his or her own performance based on personal value judgments. In order to uniformly evaluate quality of work and monitor project progress, performance standards should be established by using the Triple C approach.

Management conflict: There must be a two-way alliance between management and the project team. The views of management should be understood by the team. The views of the team should be appreciated by management. If this does not happen, management conflicts will develop. A lack of a two-way interaction can lead to strikes and industrial actions which can be detrimental to project objectives. The Triple C approach can help create a conducive dialogue environment between management and the project team.

Technical conflict: If the technical basis of a project is not sound, technical conflicts will develop. New industrial projects are particularly prone to technical conflicts because of their significant dependence on technology. Lack of a comprehensive technical feasibility study will lead to technical conflicts. Performance requirements and systems specifications can be integrated through the Triple C approach to avoid technical conflicts.

Priority conflict: Priority conflicts can develop if project objectives are not defined properly and applied uniformly across a project. Lack of a direct project definition can lead each project member to define his or her own goals which may be in conflict with the intended goal of a project. Lack of consistency of the project mission is another potential source of priority conflicts. Overassignment of responsibilities with no guidelines for relative significance levels can also lead to priority conflicts. Communication can help defuse priority conflicts.

Resource conflict: Resource allocation problems are a major source of conflict in project management. Competition for resources, including personnel, tools, hardware, and software, can lead to disruptive clashes among project members. The Triple C approach can help secure resource cooperation.

Power conflict: Project politics lead to a power play which can adversely affect the progress of a project. Project authority and project power should be clearly delineated. Project authority is the control that a person has by virtue of his or her functional post. Project power relates to the clout and influence which a person can exercise due to connections within the administrative structure. People with popular personalities can often wield a lot of project power in spite of low or nonexistent project authority. The Triple C model can facilitate a positive marriage of project authority and power to the benefit of project goals. This will help define clear leadership for a project.

Personality conflict: Personality conflict is a common problem in projects involving a large group of people. The larger a project, the larger the size of the management team needed to keep things running. Unfortunately, the larger management team creates an opportunity for personality conflicts. Communication and cooperation can help defuse personality conflicts.

In summary, conflict resolution through Triple C can be achieved by observing the following guidelines:

1. Confront the conflict and identify the underlying causes.
2. Be cooperative and receptive to negotiation as a mechanism for resolving conflicts.
3. Distinguish between proactive, inactive, and reactive behaviors in a conflict situation.
4. Use communication to defuse internal strife and competition.
5. Recognize that short-term compromise can lead to long-term gains.
6. Use coordination to work toward a unified goal.
7. Use communication and cooperation to turn a competitor into a collaborator.

MOTIVATIONAL AND INSPIRATIONAL EVALUATION

Motivation is an essential component of implementing project plans. National leaders, public employees, management staff, producers, and consumers may all need to be motivated about project plans that affect a wide spectrum of society. Those who will play active direct roles in the project must be motivated to ensure productive participation. Direct beneficiaries of the project must be motivated to make good use of the outputs of the project. Other groups must be motivated to play supporting roles to the project.

Motivation may take several forms. For projects that are of a short-term nature, motivation could be either impaired or enhanced by the strategy employed. Impairment may occur if a participant views the project as a mere disruption of regular activities or as a job without long-term benefits. Long-term projects have the advantage of giving participants enough time to readjust to the project efforts. Some of the essential considerations in aligning project plans for motivational purposes include the following elements:

Global coordination across functional lines
Balancing of task assignments
Goal-directed task analysis
Human cognitive information flow among the project team
Ergonomics and human factors considerations
Work load assessment considering fatigue, stress, emotions, sentiments, etc.
Interpersonal trust and collegiality
Project knowledge transfer lines
Harmony of personnel along project lines.

Classical concepts of motivation suggest that management involves knowing exactly what workers are expected to do and ensuring that they have the tools and skills to do it well and cost effectively. This means that management requires motivating workers to get things done. Thus, successful management should be able to predict and leverage human behavior. An effective manager should be interested in both results and the people he or she works with. Whatever definition of management is embraced, it ultimately involves some human elements with behavioral and motivational implications. In order to get a worker to work effectively, he or she must be motivated. Some workers are inherently self-motivating, self-directed, and self-actuating. There are other workers for whom motivation is an external force that must be managerially instilled based on the two basic concepts of theory X and theory Y.

THEORY X IN PROJECT EVALUATION

Theory X assumes that the worker is essentially uninterested and unmotivated to perform his or her work. Motivation must be instilled into the worker by the adoption of external motivating agents. A theory X worker is inherently indolent and requires constant supervision and prodding to get him or her to perform. To motivate a theory X worker, a mixture of managerial actions may be needed. The actions must be used judiciously, based on the prevailing circumstances. Examples of motivation approaches under theory X are as follows:

Rewards to recognize improved effort
Strict rules to constrain worker behavior
Incentives to encourage better performance
Threats to job security associated with performance failure.

THEORY Y IN PROJECT EVALUATION

Theory Y assumes that the worker is naturally interested and motivated to perform his or her job. The worker views the job function positively, and uses self-control and self-direction to pursue project goals. Under theory Y, management has the task of taking advantage of the worker's positive intuition so that his or her actions coincide with the objectives of the project. Thus, a theory Y manager attempts to use the worker's self-direction as the principal instrument for accomplishing work. In general, theory Y facilitates the following:

Worker-designed job methodology
Worker participation in decision-making
Cordial management–worker relationship
Worker individualism within acceptable company limits.

There are proponents of both theory X and theory Y, and managers who operate under each or both can be found in any organization. The important thing to note is that whatever theory one subscribes to, the approach to worker motivation should be conducive to the achievement of the overall goal of the project.

MASLOW'S HIERARCHY OF NEEDS IN PROJECT EVALUATION

The needs of project participants must be taken into consideration in any project planning in accordance with the prevailing personal and behavior landscape of the project. A common tool for accomplishing this is *Maslow's hierarchy of needs*, which stresses that human needs are ordered in a hierarchical fashion consisting of five categories:

1. *Physiological needs:* The needs for the basic things of life, such as food, water, housing, and clothing. This is the level where access to money is most critical.
2. *Safety needs:* The needs for security, stability, and freedom from threat of physical harm. The fear of adverse environmental impact may inhibit project efforts.
3. *Social needs:* The needs for social approval, friends, love, affection, and association. For example, public service projects may bring about a better economic outlook that may enable individuals to be in a better position to meet their social needs.
4. *Esteem needs:* The needs for accomplishment, respect, recognition, attention, and appreciation. These needs are important not only at the individual level but also at the national level.
5. *Self-actualization needs:* These are the needs for self-fulfillment and self-improvement. They also involve the availability of opportunity to grow professionally. Work improvement projects may lead to self-actualization opportunities for individuals to assert themselves socially and economically. Job achievement and professional recognition are two of the most important factors that lead to employee satisfaction and better motivation.

Hierarchical motivation implies that the particular motivation technique utilized for a given person should depend on where the person stands in the hierarchy of needs. For example, the need for esteem takes precedence over physiological needs when the latter are relatively well satisfied. Money, for example, cannot be expected to be a very successful motivational factor for an individual who is already on the fourth level of the hierarchy of needs. The hierarchy of needs emphasizes the fact that things that are highly craved in youth tend to assume less importance later in life.

SOCIAL HYGIENE FACTORS AND WORKFORCE MOTIVATORS

There are two motivational factors classified as the *hygiene factors* and *motivators*. Hygiene factors are necessary but not sufficient conditions for a contented worker. The negative aspects of the factors may lead to a disgruntled worker, whereas their positive aspects do not necessarily enhance the satisfaction of the worker. Examples include the following:

1. *Administrative policies:* Bad policies can lead to the discontent of workers, while good policies are viewed as routine with no specific contribution to improving worker satisfaction.
2. *Supervision:* A bad supervisor can make a worker unhappy and less productive, while a good supervisor cannot necessarily improve worker performance.
3. *Worker conditions:* Bad working conditions can enrage workers, but good working conditions do not automatically generate improved productivity.
4. *Salary:* Low salaries can make a worker unhappy, disruptive, and uncooperative, but a raise will not necessarily provoke him to perform better. While a raise in salary will not necessarily increase professionalism, a reduction in salary will most certainly have an adverse effect on morale.
5. *Personal life:* Miserable personal life can adversely affect worker performance, but a happy life does not imply that he or she will be a better worker.
6. *Interpersonal relationships:* Good peer, superior, and subordinate relationships are important to keep a worker happy and productive, but extraordinarily good relations do not guarantee that he or she will be more productive.
7. *Social and professional status:* Low status can force a worker to perform at *his* or *her* level, whereas high status does not imply performance at a higher level.
8. *Security:* A safe environment may not motivate a worker to perform better, but an unsafe condition will certainly impede productivity.

Motivators are motivating agents that should be inherent in the work itself. If necessary, work should be redesigned to include inherent motivating factors. Some guidelines for incorporating motivators into jobs are as follows:

1. *Achievement:* The job design should give consideration to opportunities for worker achievement and avenues to set personal goals to excel.
2. *Recognition:* The mechanism for recognizing superior performance should be incorporated into the job design. Opportunities for recognizing innovation should be built into the job.
3. *Work content:* The work content should be interesting enough to motivate and stimulate the creativity of the worker. The amount of work and the organization of the work should be designed to fit a worker's needs.

4. *Responsibility:* The worker should have some measure of responsibility for how his or her job is performed. Personal responsibility leads to accountability which invariably yields better work performance.
5. *Professional growth:* The work should offer an opportunity for advancement so that the worker can set his or her own achievement level for professional growth within a project plan.

The aforementioned examples may be described as job enrichment approaches with the basic philosophy that work can be made more interesting in order to induce an individual to perform better. Normally, work is regarded as an unpleasant necessity (a necessary evil). A proper design of work will encourage workers to become anxious to go to work to perform their jobs.

PROJECT MANAGEMENT BY OBJECTIVE AND EXCEPTION

Management by objective (MBO) is the management concept whereby a worker is allowed to take responsibility for the design and performance of a task under controlled conditions. It gives workers a chance to set their own objectives in achieving project goals. Workers can monitor their own progress and take corrective actions when needed without management intervention. Workers under the concept of theory Y appear to be the best suited for the MBO concept. MBO has some disadvantages which include the possible abuse of the freedom to self-direct and possible disruption of overall project coordination. The advantages of MBO include the following:

1. It encourages workers to find better ways of performing their jobs.
2. It avoids oversupervision of professionals.
3. It helps workers become better aware of what is expected of them.
4. It permits timely feedback on worker performance.

Management by exception (MBE) is an after-the-fact management approach to control. Contingency plans are not made, and there is no rigid monitoring. Deviations from expectations are viewed as exceptions to the normal course of events. When intolerable deviations from plans occur, they are investigated, and then, an action is taken. The major advantage of MBE is that it lessens the management workload and reduces the cost of management. However, it is a dangerous concept to follow especially for high-risk technology-based projects. Many of the problems that can develop in complex projects are such that after-the-fact corrections are expensive or even impossible. As a result, MBE should be carefully evaluated before adopting it. The previously described motivational concepts can be implemented successfully for specific large projects. They may be used as single approaches or in a combined strategy. The motivation approaches may be directed at individuals or groups of individuals, locally or at the national level.

PROJECT EVALUATION VIA CONSENSUS

Consensus, or lack thereof, can be a good part of project evaluation. A classic case example of project evaluation is demonstrated by the *Abilene Paradox*, which is narrated as follows. It was a July afternoon in Coleman, a tiny Texas town. It was a hot afternoon. The wind was blowing fine-grained West Texas topsoil through the house. Despite the harsh weather, the afternoon was still tolerable and potentially enjoyable. There was a fan blowing on the back porch; there was cold lemonade; and finally, there was entertainment: dominoes. Perfect for the conditions. The game required little more physical exertion than an occasional mumbled comment, "Shuffle 'em," and an unhurried movement of the arm to place the spots in the appropriate position on the table. All in all, it had the makings of an agreeable Sunday afternoon in Coleman until Jerry's father-in-law suddenly said, "Let's get in the car and go to Abilene and have dinner at the cafeteria."

Jerry thought, "What, go to Abilene? Fifty-three miles? In this dust storm and heat? And in a non-air-conditioned 1958 Buick?" But Jerry's wife chimed in with, "Sounds like a great idea. I'd like to go. How about you, Jerry?" Since Jerry's own preferences were obviously out of step with the rest, he replied, "Sounds good to me," and added, "I just hope your mother wants to go."

"Of course I want to go," said Jerry's mother-in-law. "I haven't been to Abilene in a long time." So into the car and off to Abilene they went. Jerry's predictions were fulfilled. The heat was brutal. The group was coated with a fine layer of dust that was cemented with perspiration by the time they arrived. The food at the cafeteria provided first-rate testimonial material for antacid commercials.

Some four hours and 106 miles later, they returned to Coleman, hot and exhausted. They sat in front of the fan for a long time in silence. Then, both to be sociable and to break the silence, Jerry said, "It was a great trip, wasn't it?" No one spoke. Finally, his father-in-law said, with some irritation,

> Well, to tell the truth, I really didn't enjoy it much and would rather have stayed here. I just went along because the three of you were so enthusiastic about going. I wouldn't have gone if you all hadn't pressured me into it.

Jerry couldn't believe what he just heard. "What do you mean, 'you all'?" he said. "Don't put me in the 'you all' group. I was delighted to be doing what we were doing. I didn't want to go. I only went to satisfy the rest of you. You're the culprits." Jerry's wife looked shocked.

> Don't call me a culprit. You and Daddy and Mama were the ones who wanted to go. I just went along to be sociable and to keep you happy. I would have had to be crazy to want to go out in heat like that.

Her father entered the conversation abruptly. "Hell!" he said. He proceeded to expand on what was already absolutely clear.

> Listen, I never wanted to go to Abilene. I just thought you might be bored. You visit so seldom, I wanted to be sure you enjoyed it. I would have preferred to play another game of dominoes and eat the leftovers in the icebox.

After the outburst of recrimination, they all sat back in silence. There they were, four reasonably sensible people who, of their own volition, had just taken a 106-mile trip across a godforsaken desert in a furnace-like temperature through a cloud-like dust storm to eat unpalatable food at a hole-in-the-wall cafeteria in Abilene, when one of them had really wanted to go. In fact, to be more accurate, they'd done just the opposite of what they wanted to do. The whole situation simply didn't make sense. It was a paradox of agreement.

This example illustrates a problem that can be found in many organizations or project environments. Organizations often take actions that totally contradict their stated goals and objectives. They do the opposite of what they really want to do. For most organizations, the adverse effects of such diversion, measured in terms of human distress and economic loss, can be immense. A family group that experiences the Abilene paradox would soon get over the distress, but for an organization engaged in a competitive market, the distress may last a very long time. Six specific symptoms of the paradox are identified as follows:

1. Organization members agree privately, as individuals, as to the nature of the situation or problem facing the organization.
2. Organization members agree privately, as individuals, as to the steps that would be required to cope with the situation or solve the problem they face.
3. Organization members fail to accurately communicate their desires and/or beliefs to one another. In fact, they do just the opposite and, thereby, lead one another into misinterpreting the intentions of others. They misperceive the collective reality. Members often communicate inaccurate data (e.g., "Yes, I agree"; "I see no problem with that"; "I support it") to other members of the organization. No one wants to be the only dissenting voice in the group.
4. With such invalid and inaccurate information, organization members make collective decisions that lead them to take actions

contrary to what they want to do and, thereby, arrive at results that are counterproductive to the organization's intent and purposes. For example, the Abilene group went to Abilene when it preferred to do something else.

5. As a result of taking actions that are counterproductive, organization members experience frustration, anger, irritation, and dissatisfaction with their organization. They form subgroups with supposedly trusted individuals and blame other subgroups for the organization's problems.

6. The cycle of the Abilene paradox repeats itself with increasing intensity if the organization members do not learn to manage their agreement.

This author has witnessed many project situations where, in private conversations, individuals express their discontent about a project and yet fail to repeat their statements in a group setting. Consequently, other members are never aware of the dissenting opinions. In large organizations, the Triple C model, considering the individual needs of all subsystems, can help in managing communication, cooperation, and coordination functions to avoid the Abilene paradox. The lessons to be learned from proper approaches to project planning can help avoid unwilling trips to Abilene.

CONCLUSIONS

Chapter 3 has presented diverse measures and approaches for evaluating a project. A major methodology of the chapter is the Triple C model of project communication, cooperation, and coordination. Also addressed is a quantitative measure of complexity of communication and cooperation as a function of the number of participants. The more people are involved, the more complex the pursuit of communication and cooperation.

REFERENCE

Badiru, A. B. (2008). *Triple C model of project management: Communication, cooperation, and coordination.* Boca Raton, FL: Taylor & Francis Group/CRC Press.

Project
Justification

4

*Every project failure represents an
acquisition of experience and justification to
get us ready for the next success.*

QUALITATIVE JUSTIFICATION

In a conventional sense, we often think of justification in terms of quantitative and measurable aspect. It is not necessarily so, in all cases. A project can be justified on the basis of several alternate factors, attributes, and characteristics, many of which may not be tangible or measurable. Emotions and sentiments do enter into project justification. The key is to try and achieve a hybrid measure such that subjective bias does not become the only basis for justifying a project. Since we thrive on that which we can see and measure, this chapter presents a mixture of both qualitative and quantitative factors of assessment. In any case, a project is justified on the basis of its perceived or measured value. A project can be justified on the basis of the availability of control measures and accountability framework, as noted in Figure 4.1.

QUANTITATIVE JUSTIFICATION

Cost and value are often convenient measures for justifying a project. While cost has a measurable basis, value is more elusive in a practical justification of the feasibility of a project. The feasibility of a project can be ascertained in terms of technical factors, economic factors, or both. A feasibility study is documented with a report showing all the ramifications of the project.

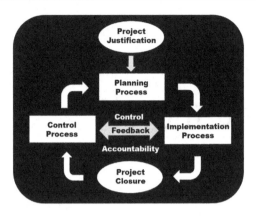

FIGURE 4.1 Control and accountability framework for project justification.

Technical feasibility: Technical feasibility refers to the ability of the process to take advantage of the current state of the technology in pursuing further improvement. The technical capability of the personnel as well as the capability of the available technology should be considered.

Managerial feasibility: Managerial feasibility involves the capability of the infrastructure of a process to achieve and sustain process improvement. Management support, employee involvement, and commitment are key elements required to ascertain managerial feasibility.

Economic feasibility: This involves the feasibility of the proposed project to generate economic benefits. A benefit–cost analysis and a break-even analysis are important aspects of evaluating the economic feasibility of new industrial projects. The tangible and intangible aspects of a project should be translated into economic terms to facilitate a consistent basis for evaluation.

Financial feasibility: Financial feasibility should be distinguished from economic feasibility. Financial feasibility involves the capability of the project organization to raise the appropriate funds needed to implement the proposed project. Project financing can be a major obstacle in large multiparty projects because of the level of capital required. Loan availability, credit worthiness, equity, and loan schedule are important aspects of financial feasibility analysis.

Cultural feasibility: Cultural feasibility deals with the compatibility of the proposed project with the cultural setup of the project environment. In labor-intensive projects, planned functions must be integrated with local cultural practices and beliefs. For example, religious beliefs may influence what an individual is willing to do or not to do.

Social feasibility: Social feasibility addresses the influences that a proposed project may have on the social system in the project environment. The ambient social structure may be such that certain categories of workers may be in short supply or nonexistent. The effect of the project on the social status of the project participants must be assessed to ensure compatibility. It should be recognized that workers in certain industries may have certain status symbols within the society.

Safety feasibility: Safety feasibility is another important aspect that should be considered in project planning. Safety feasibility refers to an analysis of whether the project is capable of being implemented and operated safely with minimal adverse effects on the environment. Unfortunately, environmental impact assessment is often not adequately addressed in complex projects.

Political feasibility: A politically feasible project may be referred to as a "politically correct project." Political considerations often dictate the direction for a proposed project. This is particularly true for large projects with national visibility that may have significant government inputs and political implications. For example, political necessity may be a source of support for a project regardless of the project's merits. On the other hand, worthy projects may face insurmountable opposition simply because of political factors. Political feasibility analysis requires an evaluation of the compatibility of project goals with the prevailing goals of the political system.

In general terms, the elements of a feasibility analysis for a project should cover the following items:

1. *Need analysis*: This indicates a recognition of a need for the project. The need may affect the organization itself, another organization, the public, or the government. A preliminary study is then conducted to confirm and evaluate the need. A proposal of how the need may be satisfied is then made. Pertinent questions that should be asked include the following:
   a. Is the need significant enough to justify the proposed project?
   b. Will the need still exist by the time the project is completed?
   c. What are the alternate means of satisfying the need?
   d. What are the economic, social, environmental, and political impacts of the need?
2. *Process work*: This is the preliminary analysis done to determine what will be required to satisfy the need. The work may be performed by a consultant who is an expert in the project field. The preliminary study often involves system models or prototypes. For technology-oriented projects, artist's conceptions, and scaled-down models may be used for illustrating the general characteristics of a process. A simulation of the proposed system can be carried out to predict the outcome before the actual project starts.
3. *Engineering and design*: This involves a detailed technical study of the proposed project. Written quotations are obtained from suppliers and subcontractors as needed. Technological capabilities are evaluated as needed. Product design, if needed, should be done at this stage.

4. *Cost estimate*: This involves estimating project cost to an acceptable level of accuracy. Levels of around −5% to +15% are common at this stage of a project plan. Both the initial and operating costs are included in the cost estimation. Estimates of capital investment and recurring and nonrecurring costs should also be contained in the cost estimate document. Sensitivity analysis can be carried out on the estimated cost values to see how sensitive the project plan is to the estimated cost values.

5. *Financial analysis*: This involves an analysis of the cash flow profile of the project. The analysis should consider rates of return, inflation, sources of capital, payback periods, break-even point, residual values, and sensitivity. This is a critical analysis since it determines whether or not and when funds will be available to the project. The project cash flow profile helps support the economic and financial feasibility of the project.

6. *Project impacts*: This portion of the feasibility study provides an assessment of the impact of the proposed project. Environmental, social, cultural, political, and economic impacts may be some of the factors that will determine how a project is perceived by the public. The value-added potential of the project should also be assessed. A value-added tax may be assessed based on the price of a product and the cost of the raw material used in making the product. A tax so collected may be viewed as a contribution to government coffers.

7. *Conclusions and recommendations*: The feasibility study should end with the overall outcome of the project analysis. This may indicate an endorsement or disapproval of the project. Recommendations on what should be done should be included in this section of the feasibility study.

JUSTIFICATION VIA PROJECT PROPOSAL

Once a project is shown to be feasible, the next step is to issue a *request for proposal* (RFP) depending on the funding sources involved. Proposals are classified as either "solicited" or "unsolicited." Solicited proposals are those written in response to a request for a proposal, while unsolicited ones are those written without a formal invitation from the funding source. Many companies prepare proposals in response to inquiries received from potential clients. Many proposals are written under competitive bids. If an RFP is issued, it should include statements about project scope, funding level, performance criteria, and deadlines.

The purpose of the RFP is to identify companies that are qualified to successfully conduct the project in a cost-effective manner. Formal RFPs are sometimes issued to only a selected list of bidders who have been preliminarily evaluated as being qualified. These may be referred to as *targeted* RFPs. In some cases, general or open RFPs are issued and whoever is interested may bid for the project. This, however, has been found to be inefficient in many respects. Ambitious, but unqualified, organizations waste valuable time preparing losing proposals. The receiving agency, on the other hand, spends much time reviewing and rejecting worthless proposals. Open proposals do have proponents who praise their "equal opportunity" approach.

In industry, each organization has its own RFP format, content, and procedures. The request is called by different names including PI (procurement invitation), PR (procurement request), RFB (request for bid), or IFB (invitation for bids). In some countries, it is sometimes referred to as request for tender (RFT). Irrespective of the format used, an RFP should request information on bidder's costs, technical capability, management, and other characteristics. It should, in turn, furnish sufficient information on the expected work. A typical detailed RFP should include

1. *Project background*: Need, scope, and preliminary studies and results.
2. *Project deliverables and deadlines*: What products are expected from the project, when the products are expected, and how the products will be delivered, should be contained in this document.
3. *Project performance specifications*: Sometimes, it may be more advisable to specify system requirements rather than rigid specifications. This gives the system or project analysts the flexibility to utilize the most updated and most cost-effective technology in meeting the requirements. If rigid specifications are given, what is specified is what will be provided regardless of the cost and the level of efficiency.
4. *Funding level*: This is sometimes not specified because of nondisclosure policies of because of budget uncertainties. However, whenever possible, the funding level should be indicated in the RFP.
5. *Reporting requirements*: Project reviews, format, number and frequency of written reports, oral communication, financial disclosure, and other requirements should be specified.
6. *Contract administration*: Guidelines for data management, proprietary work, progress monitoring, proposal evaluation procedure, requirements for inventions, trade secrets, copyrights, and so on should be included in the RFP.

7. *Special requirements (as applicable)*: Facility access restrictions, equal opportunity/affirmative actions, small business support, access facilities for the handicapped, false statement penalties, cost sharing, compliance with government regulations, and so on should be included if applicable.

8. *Boilerplates (as applicable)*: There are special requirements that specify the specific ways certain project items are handled. Boilerplates are usually written based on organizational policy and are not normally subject to conditional changes. For example, an organization may have a policy that requires that no more than 50% of a contract award will be paid prior to the completion of the contract. Boilerplates are quite common in government-related projects. Thus, large projects may need boilerplates dealing with environmental impacts, social contributions, and financial requirements.

Whether responding to an RFP or preparing an unsolicited proposal, a proposing organization must take care to provide enough detail to permit an accurate assessment of a project proposal. The proposing organization will need to find out the following:

Project time frame
Level of competition
Organization's available budget
Organization of the agency
Person to contact within the agency
Previous contracts awarded by the agency
Exact procedures used in awarding contracts
Nature of the work done by the funding agency.

The proposal should present the detailed plan for executing the proposed project. The proposal may be directed to a management team within the same organization or to an external organization. However, the same level of professional preparation should be practiced for both internal and external proposals. The proposal contents may be written in two parts: technical section and management section.

1. Technical section of project proposal
   a. Project background
      i. Expertise in the project area
      ii. Project scope
      iii. Primary objectives
      iv. Secondary objectives

   b. Technical approach
      i. Required technology
      ii. Available technology
      iii. Problems and their resolutions
      iv. Work breakdown structure
   c. Work statement
      i. Task definitions and list
      ii. Expectations
   d. Schedule
      i. Gantt charts
      ii. Milestones
      iii. Deadlines
   e. Project deliverables
   f. Value of the project
      i. Significance
      ii. Benefit
      iii. Impact
2. Management section of project proposal
   a. Project staff and experience
      i. Staff vita
   b. Organization
      i. Task assignment
      ii. Project manager, liaison, assistants, consultants, and so on
   c. Cost analysis
      i. Personnel cost
      ii. Equipment and materials
      iii. Computing cost
      iv. Travel
      v. Documentation preparation
      vi. Cost sharing
      vii. Facilities cost
   d. Delivery dates
      i. Specified deliverables
   e. Quality control measures
      i. Re-work policy
   f. Progress and performance monitoring
      i. Productivity measurement
   g. Cost control measures.

An executive summary or cover letter may accompany the proposal. The summary should briefly state the capability of the proposing organization in terms of previous experience on similar projects, unique qualification of the project personnel, advantages of the organization over other potential bidders, and reasons why the project should be awarded to the bidder.

In some cases, it may be possible to include an incentive clause in a proposal in an attempt to entice the funding organization. An example is the use of cost sharing arrangements. Other frequently used project proposal incentives include bonus and penalty clauses, employment of minorities, public service, and contribution to charity. If incentives are allowed in project proposals, their nature should be critically reviewed. If not controlled, a project incentive arrangement may turn out to be an opportunity for an organization to buy itself into a project contract.

PROJECT BUDGET JUSTIFICATION

After the planning for a project has been completed, the next step is the allocation of the resources required to implement the project plan. This is referred to as budgeting or capital rationing. Budgeting is the process of allocating scarce resources to the various endeavors of an organization. It involves the selection of a preferred subset of a set of acceptable projects due to overall budget constraints. Budget constraints may result from restrictions on capital expenditures, shortage of skilled personnel, shortage of materials, or mutually exclusive projects. The budgeting approach can be used to express the overall organizational policy. The budget serves many useful purposes including the following:

Performance measure
Incentive for efficiency
Project selection criterion
Expression of organizational policy
Plan of resource expenditure
Catalyst for productivity improvement
Control basis for managers and administrators
Standardization of operations within a given horizon.

The preliminary effort in the preparation of a budget is the collection and proper organization of relevant data. The preparation of a budget for a project is more difficult than the preparation of budgets for regular and permanent organizational endeavors. Recurring endeavors usually generate historical data which serve as inputs to subsequent estimating functions. Projects, on the other hand, are often onetime undertakings without the benefits of prior data. The input data for the budgeting process may include inflationary trends, cost of capital, standard cost guides, past records, and forecast projections. Budget data collection may be accomplished by one of several available approaches including top-down budgeting and bottom-up budgeting.

TOP-DOWN BUDGETING

Top-down budgeting involves collecting data from upper-level sources such as top and middle managers. The cost estimates supplied by the managers may come from their judgments, past experiences, or past data on similar project activities. The cost estimates are passed to lower-level managers, who then break the estimates down into specific work components within the project. These estimates may, in turn, be given to line managers, supervisors, and so on to continue the process. At the end, individual activity costs are developed. The top management issues the global budget, while the line worker generates specific activity budget requirements.

One advantage of the top-down budgeting approach is that individual work elements need not be identified prior to approving the overall project budget. Another advantage of the approach is that the aggregate or overall project budget can be reasonably accurate, even though specific activity costs may contain substantial errors. There is, consequently, a keen competition among lower-level managers to get the biggest slice of the budget pie.

BOTTOM-UP BUDGETING

Bottom-up budgeting is the reverse of top-down budgeting. In this method, elemental activities, their schedules, descriptions, and labor skill requirements are used to construct detailed budget requests. The line workers who are actually performing the activities are requested to furnish cost estimates. Estimates are made for each activity in terms of labor time, materials, and machine time. The estimates are then converted to dollar values. The dollar estimates are combined into composite budgets at each successive level up the budgeting hierarchy. If estimate discrepancies develop, they can be resolved through intervention to senior management, junior management, functional managers, project managers, accountants, or financial consultants. Analytical tools such as learning curve analysis, work sampling, and statistical estimation may be used in the budgeting process as appropriate to improve the quality of cost estimates.

All component costs and departmental budgets are combined into an overall budget and sent to top management for approval. A common problem with bottom-up budgeting is that individuals tend to overstate their needs with the notion that top management may cut the budget by some percentage. It should be noted, however, that sending erroneous and misleading estimates will only lead to a loss of credibility. Properly documented and justified budget requests are often spared the budget ax. Honesty and accuracy are invariably the best policies for budgeting.

ZERO-BASE BUDGETING

Zero-base budgeting is a budgeting approach that bases the level of project funding on previous performance. It is normally applicable to recurring programs especially in the public sector. Accomplishments in past funding cycles are weighed against the level of resource expenditure. Programs that are stagnant in terms of their accomplishments relative to budget size do not receive additional budgets. Programs that have suffered decreasing yields are subjected to budget cuts or even elimination. On the other hand, programs that experience increments in accomplishments are rewarded with larger budgets.

A major problem with zero-base budgeting is that it puts participants under tremendous data collection, organization, and program justification pressures. Too much time may be spent documenting program accomplishments to the extent that productivity improvement on current projects may be sacrificed. For this reason, the approach has received only limited use in practice. However, proponents believe it is a good means of making managers and administrators more conscious of their management responsibilities. In a project management context, the zero-base budgeting approach may be used to eliminate specific activities that have not contributed to project goals in the past.

PROJECT JUSTIFICATION BY VALUE MODELING

A technique that can be used to assess overall value-added components of a process improvement program is the systems value model (SVM). The model provides an analytical decision aid for comparing process alternatives. Value is represented as a p-dimensional vector:

$$V = f\left(A_1, A_2, \ldots, A_p\right)$$

where $A = \left(A_1, \ldots, A_n\right)$ is a vector of quantitative measures of tangible and intangible attributes. Examples of process attributes are quality, throughput, capability, productivity, cost, and schedule. Attributes are considered to be a combined function of factors, x_1, expressed as follows:

$$A_k\left(x_1, x_2, \ldots, x_{m_k}\right) = \sum_{i=1}^{m_k} f_i\left(x_i\right)$$

where $\{x_i\}$ = set of m factors associated with attribute A_k $(k = 1, 2, \ldots, p)$ and f_i = contribution function of factor x_i to attribute A_k. Examples of factors include reliability, flexibility, user acceptance, capacity utilization, safety, and design functionality. Factors are themselves considered to be composed of indicators, v_i, expressed as

$$x_i\left(v_1, v_2, \ldots, v_n\right) = \sum_{j=1}^{n} z_i\left(v_i\right)$$

where $\{v_j\}$ = set of n indicators associated with factor x_i $(i = 1, 2, \ldots, m)$, and z_j = scaling function for each indicator variable v_j. Examples of indicators are project responsiveness, lead time, learning curve, and work rejects. By combining the above definitions, a composite measure of the value of a process can be modeled as follows:

$$V = f\left(A_1, A_2, \ldots, A_p\right)$$

$$= f\left\{\left[\sum_{i=1}^{m_1} f_i\left(\sum_{j=1}^{n} z_j\left(v_j\right)\right)\right]_1, \left[\sum_{i=1}^{m_2} f_i\left(\sum_{j=1}^{n} z_j\left(v_j\right)\right)\right]_2, \ldots, \left[\sum_{i=1}^{m_k} f_i\left(\sum_{j=1}^{n} z_j\left(v_j\right)\right)\right]_p\right\}$$

where m and n may assume different values for each attribute. A subjective measure to indicate the utility of the decision-maker may be included in the model by using an attribute weighting factor, w_i, to obtain a weighted PV:

$$PV_w = f\left(w_1 A_1, w_2 A_2, \ldots, w_p A_p\right)$$

where

$$\sum_{k=1}^{p} w_k = 1, \qquad (0 \le w_k \le 1)$$

With this modeling approach, a set of process options can be compared on the basis of a set of attributes and factors.

To illustrate the model above, suppose three IT options are to be evaluated based on four attribute elements: *capability, suitability, performance,* and *productivity.* For this example, based on the equations, the value vector is defined as follows:

$$V = f\left(capability, suitability, performance, productivity\right)$$

Capability: The term "capability" refers to the ability of IT equipment to satisfy multiple requirements. For example, a certain piece of IT equipment may only provide computational service. A different piece of equipment may be capable of generating reports in addition to computational analysis, thus increasing the service variety that can be obtained. In the analysis, the levels of increase in service variety from the three competing equipment types are 38%, 40%, and 33%, respectively. *Suitability*: "Suitability" refers to the appropriateness of the IT equipment for current operations. For example, the respective percentages of operating scope for which the three options are suitable for are 12%, 30%, and 53%. *Performance*: "Performance," in this context, refers to the ability of the IT equipment to satisfy schedule and cost requirements. In the example, the three options can, respectively, satisfy requirements on 18%, 28%, and 52% of the typical set of jobs. *Productivity*: "Productivity" can be measured by an assessment of the performance of the proposed IT equipment to meet workload requirements in relation to the existing equipment. For the example, the three options, respectively, show normalized increases of 0.02, −1.0, and −1.1 on a uniform scale of productivity measurement. Option C is the best "value" alternative in terms of suitability and performance. Option B shows the best capability measure, but its productivity is too low to justify the needed investment. Option A offers the best productivity, but its suitability measure is low. The analytical process can incorporate a lower control limit into the quantitative assessment such that any option providing value below that point will not be acceptable. Similarly, a minimum value target can be incorporated into the graphical plot such that each option is expected to exceed the target point on the value scale.

The relative weights used in many justification methodologies are based on subjective propositions of decision-makers. Some of those subjective weights can be enhanced by the incorporation of utility models. For example, the weights could be obtained from utility functions. There is a risk of spending too much time maximizing inputs at "point-of-sale" levels with little time defining and refining outputs at the "wholesale" systems level.

A systems view of operations is essential for every organization. Without a systems view, we cannot be sure we are pursuing the right outputs. A systems approach allows for a multi-dimensional analysis of any endeavor, considering many of the typical "ilities" of systems engineering as listed below:

- Affordability
- Practicality
- Desirability
- Configurability
- Modularity
- Reliability
- Desirability
- Maintainability
- Testability
- Transmittability
- Reachability
- Quality
- Agility.

A systems engineering plan is essential for the following reasons:

1. Description of the system being developed
2. Description of team structure and responsibilities
3. Identification of all project stakeholders
4. Description of tailored technical activities in each phase
5. Documentation of decisions and technical implementation
6. Establishment of technical metrics and measurements (Who, What, When, Where, Which, How, Why).

Now that we have explained some of the characteristics of a system, we can move on to other specific applications and other considerations.

COST AND VALUE OF JUSTIFICATION INFORMATION

Information is the basis for planning. However, too much information is as bad as too little information. Too much information can impede the progress of a project. The marginal benefit of information decreases as its size increases. However, the marginal cost of obtaining additional information may increase as the size of the information increases. Figure 4.2 shows the potential behaviors of the value and cost curves with respect to the size of information.

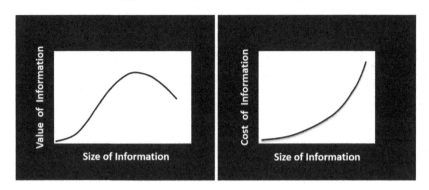

FIGURE 4.2   Plots of cost and value of information.

The optimum size of information is determined by the point that represents the widest positive difference between the value of information and its cost. The costs associated with information can often be measured accurately. However, the benefits may be more difficult to document. The size of information may be measured in terms of a number of variables including number of pages of documentation, number of lines of code, and the size of the computer storage requirement. The amount of information presented for project management purposes should be condensed to cover only what is needed to make effective decisions. Information condensation may involve pruning the information that is already available or limiting what needs to be acquired.

The cost of information is composed of the cost of the resources required to generate the information. The required resources may include computer time, personnel hours, software, and so on. Unlike the value of information, which may be difficult to evaluate, the cost of information is somewhat more tractable. However, the development of accurate cost estimates prior to actual project execution is not trivial. The degree of accuracy required in estimating the cost of information depends on the intended uses of the estimates. Cost estimate may be used as general information for control purposes. Cost estimates may also be used as general guides for planning purposes or for developing standards. The bottom-up cost estimation approach is a popular method used in practice. This method proceeds by breaking the cost structure down into its component parts. The cost of each element is then established. The sum of these individual cost elements gives an estimate of the total cost of the information.

It is important to assess the value of project information relative to its cost before deciding to acquire it. Investments for information acquisition should be evaluated just like any other capital investment. The value of information is determined by how the information is used. In project management, information has two major uses. The first use relates to the need for information to run the daily operations of a project. Resource allocation, material procurement, re-planning, rescheduling, hiring, and training are just a few of the daily functions for which information is needed. The second major use of information in project management relates to the need for information to make long-range project decisions. The value of information for such long-range decision-making is even more difficult to estimate since the future cost of not having the information today is unknown.

The classical approach for determining the value of information is based on the value of perfect information. The expected value of perfect information is the maximum expected loss due to imperfect information. Using probability analysis or other appropriate quantitative methods, the project analyst can predict what a project outcome might be if certain information is available, or not available. For example, if it is known for sure that it will rain on a certain day, a project manager might decide to alter the project schedule so that only non-weather-sensitive tasks are planned on that particular day. The value of the perfect information about the weather would then be measured in terms of what loss could have been incurred if that information were not available. The loss may be in terms of lateness penalty, labor idle time, equipment damage, or ruined work.

An experienced project manager can accurately estimate the expected losses and hence the value of the perfect information about the weather. The cost of the same information may be estimated in terms of what it would cost to consult with a weather forecaster or the cost of buying a subscription to a special weather forecast channel on cable television.

ANALYTICS FOR WORKER ASSIGNMENT

Project justification can also have a basis in the availability and assignment of workers as a resource allocation problem. Operations research techniques are frequently used to enhance resource allocation decisions. One common resource allocation tool is the resource assignment algorithm, which can be applied to enhance human resource management. Suppose that there are n tasks which must be performed by n workers. The cost of worker i performing task j is c_{ij}. It is desired to assign workers to the tasks in a fashion that minimizes the cost of completing the tasks. This problem scenario is referred to as the *assignment problem*. The technique for finding the optimal solution to the problem is called the *assignment method*. Like the transportation method, the assignment method is an iterative procedure that arrives at the optimal solution by improving on a trial solution at each stage of the procedure. Conventional critical path method (CPM) and project evaluation and review technique (PERT) can be used in controlling projects to ensure that the project will be completed on time, but both techniques do not consider the assignment of resources to the tasks that make up a project. The *assignment method* can be used to achieve an optimal assignment of resources to specific tasks in a project. Although the assignment method is cost-based, task duration can be incorporated into the modeling in terms of time–cost relationships. Of course, task precedence requirements and other scheduling constraints of the tasks will be factored into the computational procedure. The objective is to minimize the total cost of assigning workers to tasks. The formulation of the assignment problem is as follows:

Let

$x_{ij} = 1$ if worker i is assigned to task j, where $i, j = 1, 2, \ldots, n$

$x_{ij} = 0$ if worker i is not assigned to task j

$c_{ij} =$ cost of worker i performing task j

$$\text{Minimize } z = \sum_{i=1}^{n} \sum_{j=1}^{n} c_{ij} x_{ij}$$

$$\text{Subject to } \sum_{j=1}^{n} x_{ij} = 1, \quad i = 1, 2, \ldots, n$$

$$\sum_{i=1}^{n} x_{ij} = 1, \quad j = 1, 2, \ldots, n$$

$$x_{ij} \geq 0, \quad i, j = 1, 2, \ldots, n$$

The previous formulation is a transportation problem with $m = n$ and all supplies and demands (sources to targets) are equal to 1. Note that we have used the nonnegativity constraint, $x_{ij} \geq 0$, instead of the integer constraint, $x_{ij} = 0$ or 1. However, the solution of the model will still be integer valued. Hence, the assignment problem is a special case of the transportation problem with $m = n$; $S_i = 1$ (supplies); and $D_i = 1$ (demands). Conversely, the transportation problem can also be viewed as a special case of the assignment problem. The basic requirements of an assignment problem are as follows:

1. There must be two or more tasks to be completed.
2. There must be two or more resources that can be assigned to the tasks.
3. The cost of using any of the resources to perform any of the tasks must be known.
4. Each resource is to be assigned to one and only one task.

If the number of tasks to be performed is greater than the number of workers available, we will need to add *dummy workers* to balance the problem formulation. Similarly, if the number of workers is greater than the number of tasks, we will need to add *dummy tasks* to balance the formulation. If there is no problem of overlapping, a worker's time may be split into segments so that the worker can be assigned more than one task. In this case, each segment of the worker's time will be modeled as a separate resource in the assignment problem. Thus, the assignment problem can be extended to consider partial allocation of resource units to multiple tasks.

Although the assignment problem can be formulated for and solved by the simplex method or the transportation method, a more efficient algorithm is available specifically for the assignment problem. The method, known as the *Hungarian method*, is a simple iterative technique. Details of the assignment problem and its solution techniques can be found in operations research textbooks.

PROJECT JUSTIFICATION THROUGH TERMINATION ANALYSIS

Project termination is an important aspect of project control. Termination should be viewed as a control function since some projects can drag on unnecessarily if control is not instituted. There are several reasons for terminating projects. Some projects are terminated under cordial, arranged, and expected circumstances, while others are terminated under unpleasant circumstances that call for managerial control. If necessary, a project audit should be conducted to ascertain the need to terminate a project. Some of the common reasons include the following:

Cost overruns
Alternate technology
Missed deadline
Product obsolescence
Environmental concern
Government requirement
Excessive delay penalties
Technically impossible goals
Lack of project justification
Poor performance beyond remedy
Alternate objective to the initial plan

Project objective accomplished
Poor unachievable project plan
Lack of required personnel or other resources.

Even after the reasons for terminating the project have been identified, actual termination may not be easy to implement especially for long-range and large projects that have spread their tentacles throughout an organization. Problems of morale may develop. Some workers may have grown accustomed to the extra attention, recognition, or advancement opportunities associated with the project. They may not see the wisdom of terminating the project. The Triple C approach should be used in setting the stage for the termination of a project at the appropriate time. The termination process should cover the following items:

Communicating with the personnel on the need for termination
Retraining workers for new functions
Reassigning workers to other functions
Assuring the cooperation of those involved
Returning workers to their previous functions
Coordinating the required actions for termination
Withdrawing funding from the project (pulling the plug).

If the termination is handled properly, workers will be less agonized by the loss and there will be a smooth transition to other projects.

Project system control must be verified and validated before being implemented in a functional setting. Without proper verification and validation, disappointing results can occur. An important reason for performing careful verification is based on the fact that when a system malfunctions, the source of the problem may not be as obvious as in the case of conventional programs. Thus, the problem may go undetected until serious harm has been done to the project or the organization.

Verification involves the determination of whether or not the system is functioning as intended. This may involve program debugging, error analysis, input acceptance check, output verification, reasonableness of operation, run time control, and result documentation.

Validation concerns a diagnosis of how closely a system's solution matches expected solution. If the system is valid, then the decisions, conclusions, or recommendations it offers can form the basis for setting actual operating conditions. The validation should be done by using different problem scenarios to simulate actual system's operations.

Sometimes, it may be impossible to validate a system for all the anticipated problems because the data for such problems may not yet be available.

In such a case, the closest possible representation of the expected scenarios should be utilized for the validation.

The crucial characteristic components of a system should be identified and used as the basis for validation. For example, if the project involves the development of a software tool, then the knowledge base component of the software is the area which will require the most thorough evaluation since it contains the problem-solving strategies of the software.

The methodology that one uses to determine how much validation to perform on a system is dependent on the number of representative cases which are available for evaluation. For example, evaluating a medical knowledge base which diagnoses rare diseases will be much more difficult than evaluating a knowledge base which addresses a common problem of workshop layout. Based on the availability of representative cases, special techniques may need to be used to perform validation (i.e., sensitivity analysis, what-if analysis).

The degree of validation to be performed on a system is dependent on the degree of significance placed on the system. This is based on the context in which the system will be used. In some cases, a system may be viewed as a complete replacement for a human operator. This results in total dependence on the system and a need for greater validation. In other cases, a system may be developed as a complementary tool in problem-solving. This type of use does not require very rigorous validation.

The appropriate time to perform validation is a key decision in any project. Since errors can occur anywhere in the development process from data collection to methodology development, validating a system in stages is very important. Validating a system in stages facilitates catching errors before they become compounded. For very small systems, validation can be performed in one single stage at the completion of the system.

In a very large system, validation should occur at each stage of the development cycle. Small and medium systems may be validated at only a few selected intervals. Recommended stages for validation are as follows:

1. *Conceptualization stage with overall goal definition*: State what the measures of the project's success will be and how failure or success will be evaluated.
2. *First version prototype showing feasibility*: Demonstrate feasibility of the system and perform preliminary evaluation with a few special test cases.
3. *System refinement*: Evaluate with informal test cases, and get feedback from experts, end users, and other stakeholders.
4. *Evaluation of performance*: Perform formalized evaluation using randomly selected data inputs.

5. *Evaluation of acceptability to users*: Evaluate the system in its intended users' surroundings. Verify that the system has good human factors interface (i.e., input/output devices and ease of use).

6. *Evaluation of functionality for extended period in prototype environment*: Field test and verify the system. Observe performance of the system and reactions of the users.

7. *Pre-implementation evaluation*: Evaluate the overall system prior to deployment in an operating environment.

FACTORS INVOLVED IN JUSTIFICATION

Several major factors should be examined carefully in the verification and validation stages of a system for the purpose of justification. The objective is to verify and validate that for any correct input to the system, a correct output can be obtained. Factors of interest in system validation include the following:

1. *Completeness*: This refers to the thoroughness of the system and checks if the system can address all desired problems within its problem domain.

2. *Efficiency*: Efficiency checks how well the system makes use of the available knowledge, data, hardware, software, and time in solving problems within its specified domain.

3. *Validity*: This involves the correctness of the system outputs. Validity may be viewed as the ability of the system to provide accurate results for relevant data inputs.

4. *Maintainability*: This involves how well the integrity of the system can be preserved even when operating conditions change.

5. *Consistency*: Consistency requires that the system provide similar results to similar problem scenarios.

6. *Precision*: This refers to the level of certainty or reliability associated with the consultations provided by the system. Precision is often application dependent. For example, precision in a medical diagnosis may have more importance than precision in other domains of diagnosis. Compliance with any prevailing rules and regulations is an important component of the precision of a technical system.

7. *Soundness*: Soundness refers to the quality of the scientific and technical basis for the methodology of the system.

8. *Usability*: This involves an evaluation of how the system might meet users' needs. Questions to be asked include the following: Is the system usable by the end user? Are questions worded in an easily understood format? Is help available? Is the system able to explain its reasoning process to the user? Is the system compatible with the delivery environment?

9. *Justification*: A key factor of validation involves justification. A system should be justified in terms of cost requirements, operating characteristics, maintainability, and responsiveness to user requests.

10. *Reliability*: Under reliability evaluation, the system is expected to perform satisfactorily whenever it is used. It should not be subject to erratic performance and results. Several test runs are typically needed to ascertain the reliability of a system.

11. *Accommodating*: To be accommodating, the system has to be very forgiving for minor data entry errors by the user. Appropriate prompts should be incorporated into the user interface to inform users of incorrect data inputs and allow corrections of inputs.

12. *Clarity*: Clarity refers to how well the system presents its prompts to avoid ambiguities in the input/output processes. If the system possesses a high level of clarity, there will be assurance that it will be used as intended by the users.

13. *Quality*: The quality of a system refers to the subjective perception of the user of the system. Quality is often defined as a measure of the user's satisfaction. It refers to the comprehensive combination of the characteristics of a system that determines the system's ability to satisfy specific needs.

14. *Other "ilities"*: Modularity, reconfigurability, interoperability, utility, etc.

HOW TO EVALUATE THE SYSTEM

To correctly validate a system based on empirical analysis, the correct results for test cases must be known and accessible. With known results, an absolute measure of the effectiveness of the system may be estimated as the proportion of correct to incorrect results produced by the system. If standard results are not available, then a relative evaluation of the system may be performed on the basis of the performance of other systems designed to perform similar functions. Provided in the following text are some guidelines on how to perform the evaluation process:

Set realistic standards for the performance of the system.

Define the minimum acceptable standard required for the system to be considered successful.

Use performance standards that are comparable to those used in evaluating comparable systems.

Use controlled experiments whereby the evaluators are not biased by the sources of the results being evaluated.

Distinguish between "false-positive" and "true-positive" results produced by the expert system. In a false-positive result, the system would diagnose as "true" what is not really "true." In true-positive results, the system would diagnose as "true" what is really "true."

In cases of incorrect results, identify which correct solutions are closest to being reached. This will be very valuable in performing a refinement of the system later on.

SENSITIVITY ANALYSIS FOR PROJECT CONTROL

To improve the precision of a project control system, sensitivity analysis can be performed by the project team. Sensitivity analysis establishes the variability in the conclusions of the system as a function of the variability of the data. That is, we would identify the differences in results that are caused by different levels of changes in the input to the system. If minor changes in the inputs lead to large differences in the result, then the system is said to be very sensitive to changes in inputs. One effective method of using sensitivity analysis to improve precision is to display as a histogram output values against possible answers for one given input. Sensitive points will be displayed as significant changes in the histogram. These visual identifications help identify potential trouble spots in the system.

Case Study: Commentary on Post Office Justification

This case study is a community-oriented commentary by the author in response to the 2020 debate on the US postal system, its lack of profitability, and the post master general's abrupt decision to make drastic changes in the operations of the post office nationwide. The commentary represents a relevant example of justifying a project on the basis of non-quantifiable value rather than a quantitative measure of profit.

Subtle value, rather than an explicit measure, can, sometimes, be the basis for justifying a project. Let's take the case of the post office as an example. It is a delight to hear that the sense of reason has prevailed in the most recent political assault on our beloved post office system. The reversed decision of the postmaster general to defer proposed changes until after the 2020 election demonstrates where the hand of logic has touched an irrational development. The decision to make such a radical change in the post office should not have happened in the first place at this critical moment in the battle of wits with COVID-19 and the prevailing shifty political landscape. Although many community voices were raised about the decision to make post office changes, I have not seen community-wide placards declaring "**Don't mess with our post office.**" As time goes on, I hope more of the community will be sensitized to the plight of the post office, which is often lambasted for inefficiency, slowness, and unprofitability. Beyond just processing mails from and to desired points within the community, many people don't realize the expansive value that the post office brings to the community. In any corporate setting, there are profit-generating entities just as there are non-profits. To constantly evaluate the post office on the basis of operational financial losses alone is to ignore the larger value of the post office as a community rallying point. The premise of establishing the post office in the first place was what was envisioned as the need for a community connection. Since its inception in 1775, the post office has dedicatedly provided that value. Whenever we consider the efficiency or profitability of the post office, we should always, concurrently, evaluate its value, with respect to the growing services to a rapidly growing and diverse population. This lack of value appreciation is where I often see a disconnect in the reactions to the lack of profitability of the post office. Many people don't realize that the post office is self-sustaining. Without a dedicated national funding, the post office will always be fighting an uphill battle. Certainly, there is always room for improvement in any organization, and the post has been doing its best. In my opinion, even though the losses may be mounting, I believe the value to the community is actually increasing. Unfortunately, value is not readily quantifiable. Thus, observers latch on to that which they can see and measure in terms of post office losses.

In more and more deep thinking about this issue, I have discovered that the reason I love going to the post office so much is the same reason that I am addicted to maintaining my consistent subscription to local newspapers wherever I have lived for over 45 years. That common reason is the community connectedness that the post office and the local newspaper provide. No matter how digital we take our operations, humans will still need human interaction and interfaces. Post office and local newspapers represent the consistent and reliable avenues for us to continue to be humans.

Telephone booths used to be a symbol of the community and a rallying point for all sorts of communications, both wholesome and illicit. But the advent of cell phones spelled doom for the phone booths, which are rarely seen these days. This should not be allowed to happen to the post office and the local newspaper. Community mail boxes should not be removed! Personally, I detest the recent promotional Ads by the post office itself, encouraging customers to go digital and "never visit the post office again." Hogwash! That's like inviting us to stop being humans. The post office is the remaining, but thinning, thread of human connectivity in the community. It should not be allowed to be assaulted unabashedly by any political leanings. Congress should fund the post office and help to increase its value to the community rather than letting it wither in the wilderness of modern sentiments of going digital. This is an argument for the justification of the existence and continuation of human-centered services at the post office.

CONCLUSIONS

Chapter 4 has presented alternate tools and techniques for project justification. The key proposition is that a project can be justified not only on quantifiable metrics but also on the basis of qualitative value. This chapter also introduces plots of the cost and value of information with respect to the size of information.

REFERENCE

Badiru, A. B. (2019). *Project management: Systems, principles, and applications* (2nd ed.), Boca Raton, FL: Taylor & Francis Group/CRC Press.

Project Integration

5

*Project lessons learned should be project
lessons practiced and integrated.*
– Adedeji Badiru

IMPORTANCE OF PROJECT INTEGRATION

Integration is a key aspect of project implementation. An implementation that is not properly integrated into the operating culture and infrastructure of the organization is bound to fail.

Systems integration permits sharing of resources. Physical equipment, concepts, information, and skills may be shared as resources. Systems integration is now a major concern of many organizations. Even some of the organizations that traditionally compete and typically shun cooperative efforts are beginning to appreciate the value of integrating their operations. For these reasons, systems integration has emerged as a major interest in business. Systems integration may involve the physical integration of technical components, objective integration of operations, conceptual integration of management processes, or a combination of any of these.

Systems integration involves the linking of components to form subsystems and the linking of subsystems to form composite systems within a single department and/or across departments. It facilitates the coordination of technical and managerial efforts to enhance organizational functions, reduce cost, save energy, improve productivity, and increase the utilization of resources. Systems integration emphasizes the identification and coordination of the interface requirements among the components in an integrated system. The components and subsystems operate synergistically to optimize the performance of the total system. Systems integration ensures that all performance goals are satisfied with a minimum expenditure of time and resources. Integration can be achieved in several forms including the following:

1. *Dual-use integration*: This involves the use of a single component by separate subsystems to reduce both the initial cost and the operating cost during the project life cycle.
2. *Dynamic resource integration*: This involves integrating the resource flows of two normally separate subsystems so that the resource flow from one to or through the other minimizes the total resource requirements in a project.
3. *Restructuring of functions*: This involves the restructuring of functions and reintegration of subsystems to optimize costs when a new subsystem is introduced into the project environment.

Systems integration is particularly important when introducing new technology into an existing system. It involves coordinating new operations to coexist with existing operations. It may require the adjustment of functions to permit the sharing of resources, development of new policies to accommodate product integration, or realignment of managerial responsibilities. It can affect both hardware and software components of an organization. Presented in the following list are guidelines and important questions relevant for systems integration:

What are the unique characteristics of each component in the integrated system?
How do the characteristics complement one another?
What physical interfaces exist among the components?
What data/information interfaces exist among the components?
What ideological differences exist among the components?
What are the data flow requirements for the components?
Are there similar integrated systems operating elsewhere?
What are the reporting requirements in the integrated system?
Are there any hierarchical restrictions on the operations of the components of the integrated system?
What internal and external factors are expected to influence the integrated system?
How can the performance of the integrated system be measured?
What benefit/cost documentations are required for the integrated system?
What is the cost of designing and implementing the integrated system?
What are the relative priorities assigned to each component of the integrated system?

What are the strengths of the integrated system?

What are the weaknesses of the integrated system?

What resources are needed to keep the integrated system operating satisfactorily?

Which section of the organization will have primary responsibility for the operation of the integrated system?

What are the quality specifications and requirements for the integrated systems?

PROJECT MANAGEMENT STEPS PLUS INTEGRATION

The conventional steps of project management are illustrated in Figure 5.1. In accordance with the methodology of the DEJI System Model, the step of integration has been added.

The end result of every project should be an integration into the existing or evolving work environment. An alignment with workforce needs and what the organization desires is essential for having a sustainable utilization of the project outcome. For example, the conventional processes of leveraging Maslow's hierarchy of needs of humans will likely be modified due to the evolving work environment of the future. A systems-based integration,

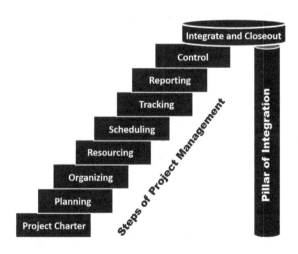

FIGURE 5.1 Steps of project management capped with project integration.

as advocated by the DEJI Systems Model, will be needed for all projects. Of what use is a productive project that is out of phase with the existing work scenarios and preferences. For example, the classical processes of CPM (critical path method) and PERT (project evaluation and review technique) in moving work from point A to point B must adapt to a modern critical path. What that modern critical path looks like is a subject of a robust systems approach.

SMART INTEGRATION

The principle of smart operations using the technique of SMART (specific, measurable, aligned, realistic, and timed) for performance measurement is useful for creating a foundation for project integration. Figure 5.2 illustrates a comprehensive view of the pursuit of integration under the platform of SMART. The figure demonstrates that all existing management tools and techniques can find a place in the integration effort from the project conceptualization stage through the desired integration phase of the DEJI Systems Model. The elements of the SMART approach are summarized below:

1. *Specific*: Pursue specific and explicit outputs. The simpler a goal (in smaller chunks), the easier it will be to accomplish.
2. *Measurable*: Design of outputs that can be tracked, measured, and assessed.
3. *Achievable*: Make outputs to be achievable and aligned with organizational goals. Goals should be aligned to the prevailing operations to pave the way for agility in the integration phase.

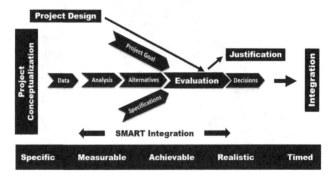

FIGURE 5.2   Framework for SMART integration.

4. *Realistic*: Pursue only the goals that are realistic and result oriented. This will facilitate resilience.

5. *Timed*: Make outputs timed to facilitate accountability and traceability.

MATHEMATICAL INTEGRATION

We must integrate all the elements of a project on the basis of alignment of functional goals. Mathematically, systems overlap for integration purposes can conceptually be represented as projection integrals by considering areas bounded by the common elements of subsystems:

$$A = \int_{A_y A_x} z(x,y)\, dy\, dx$$

$$B = \int_{B_y B_x} z(x,y)\, dy\, dx$$

If we think conceptually about two flat surfaces (planes) that are desired to be integrated, the projection of one flat plane onto the other is represented as an area of overlap. If both surfaces are perfectly flat horizontally, then the area of overlap can easily be identified and computed. The overlapping area can be defined as a measure of the level of alignment and integration. Now, consider a case where one of the surfaces is at an inclined angle. In that case, the projection of that surface onto the other will represent a smaller measure. In that case, mathematically, the net projection can then be represented as follows:

$$C = \int_{C_y C_x} z(x,y)\, dy\, dx$$

Notice how each successful net projection area decreases with an increase in the angle of inclination of the project plane. The fact is that in actual project execution, it will be impractical or impossible to model subsystem scenarios as double integrals. But the concept, nonetheless, demonstrates the need to consider where and how project elements overlap for a proper assessment of integration. For mechanical and electrical systems, one can very well develop mathematical representation of systems overlap and integration boundaries. For the purpose of mathematical exposition, double integrals arise in several technical applications. Some examples are as follows:

Calculation of volumes
Calculation of the surface area of a two-dimensional surface (e.g., a plane surface)
Calculation of a force acting on a two-dimensional surface
Calculation of the average of a function
Calculation of the mass or moment of inertia of a body
Consider the surface area given by the integral

$$A(x) = \int_c^d f(x,y)dy$$

The variable of integration is y, and x is considered a constant. The cross-sectional area depends on x. Thus, the area is a function of x. That is, $A(x)$. The volume of the slice between x and $x + dx$ is $A(x)dx$. The total volume is the sum of the volumes of all the slices between $x = a$ and $x = b$. That is,

$$V = \int_a^b A(x)dx$$

If substitute for $A(x)$, we obtain

$$V = \int_a^b \left[\int_c^d f(x,y)dy \right] dx = \int_a^b \int_c^d f(x,y)dy\,dx$$

This is an example of an *iterated* integral. One integrates with respect to y first, and then x. The integrals with respect to y and x are called the inner and outer integrals, respectively. Alternatively, one can make slices that are parallel to the x-axis. In this case, the volume is given by

$$V = \int_c^d \left[\int_a^b f(x,y)dx \right] dy = \int_c^d \int_a^b f(x,y)dx\,dy$$

The inner integral corresponds to the cross-sectional area of a slice between y and $y + dy$. The quantities $f(x, y)\,dy\,dx$ and $f(x, y)\,dx\,dy$ represent the value of the double integral in the infinitesimally small rectangle between x and $x + dx$ and

y and $y + dy$. The length and width of the rectangle are dx and dy, respectively. Hence $dy\,dx$ (or $dx\,dy$) is the area of the rectangle. Thus, the change in area is $dA = dy\,dx$ or $dA = dx\,dy$.

Consider the double integral below for a hypothetical computational example:

$$V = \iint\limits_{R} \left(x^2 + xy^3\right) dA$$

where R is the rectangle $0 \le x \le 1$, $1 \le y \le 2$. Suppose we integrate with respect to y first. Then,

$$V = \int\limits_{0}^{1}\int\limits_{1}^{2} \left(x^2 + xy^3\right) dy\,dx$$

The inner integral is

$$V = \int\limits_{1}^{2} \left(x^2 + xy^3\right) dy = \left[x^2 y + x\frac{y^4}{4}\right]_{y=1}^{y=2}$$

Note that we treat x as a constant as we integrate with respect to y. The integral is equal to

$$x^2(2) + x\left(\frac{2^4}{4}\right) - x^2 - \frac{x}{4} = x^2 + \left(\frac{15}{4}\right)x$$

We are now left with the following integral:

$$\int\limits_{0}^{1}\left(x^2 + \frac{15}{4}x\right)dx = \left(\frac{x^3}{3} + \frac{15}{8}x^2\right)_{x=0}^{x=1} = \frac{1}{3} + \frac{15}{8} = 2.2083$$

Alternatively, we can integrate with respect to x first and then y. We have

$$V = \int\limits_{1}^{2}\int\limits_{0}^{1} \left(x^2 + xy^3\right) dx\,dy$$

which should yield the same computational result.

In terms of a summary for this chapter, systems integration is the synergistic linking together of the various components, elements, and subsystems of a system, where the system may be a complex project, a large endeavor, or an expansive organization. Activities that are resident within the system must be managed from both the technical and managerial standpoints. Any weak link in the system, no matter how small, can be the reason that the overall system fails. In this regard, every component of a project is a critical element that must be nurtured and controlled. Embracing the systems principles for project management will increase the likelihood of success of projects.

MANAGEMENT PRINCIPLES FOR INTEGRATION

Project integration is a managerial pursuit that can benefit from the application of conventional management principles, such as Lean and Six Sigma. The term, "Lean," is a terminology that is well known and defined as an elimination of waste in operations through managerial principles. Many principles are comprised in the Lean concept, but the major thought to remember is effective utilization of resources and time in order to achieve higher quality products and ensure customer satisfaction. Remembering back, defects are anything that the customer is unhappy with and is a term utilized in Six Sigma. Six Sigma identifies and eliminates these defects so that the customer in turn is satisfied. The customer is the number one focus, and if they are unhappy, they will have no problem going elsewhere which most likely is a competition for the business. Coupling Lean and Six Sigma will reduce waste and reduce defects. The concept will be called Lean Six Sigma going further.

The most basic concept when discussing waste reduction begins with Kaizen. Kaizen is a Japanese concept defined as "taking apart and making better." The concept takes a vast amount of project management techniques to facilitate the process going forward. 5s processes are the most predominant and commonly known for Kaizen events.

5s principles are determined by finding a place for everything and everything in its place.

The 5s levels are as follows:

Sort: Identify and eliminate necessary items and dispose of unneeded materials that do not belong in an area. This reduces waste, creates a safer work area, opens space, and helps visualize processes. It is important to sort through the entire area. The removal of items should be discussed with all personnel involved. Items that cannot be removed immediately should be tagged for subsequent removal.

Sweep: Clean the area so that it looks like new, and clean it continuously. Sweeping prevents an area from getting dirty in the first place and eliminates further cleaning. A clean work place indicates high standards of quality and good process controls. Sweeping should eliminate dirt, build pride in work areas, and build value in equipment.

Straighten: Have a place for everything and everything in its place. Arranging all necessary items is the first step. It shows what items are required and what items are not in place. Straightening aids efficiency; items can be found more quickly and employees travel shorter distances. Items that are used together should be kept together. Labels, floor markings, signs, tape, and shadowed outlines can be used to identify materials. Shared items can be kept at a central location to eliminate purchasing more than needed.

Schedule: Assign responsibilities and due dates to actions. Scheduling guides sorting, sweeping, and straightening and prevents regressing to unclean or disorganized conditions. Items are returned where they belong and routine cleaning eliminates the need for special cleaning projects. Scheduling requires checklists and schedules to maintain and improve neatness.

Sustain: Establish ways to ensure maintenance of manufacturing or process improvements. Sustaining maintains discipline (Badiru and Kovach, 2012). Utilizing proper processes will eventually become routine. Training is key to sustaining the effort and involvement of all parties. Management must mandate the commitment to housekeeping for this process to be successful.

The benefits of 5s include (a) a cleaner and safer workplace; (b) customer satisfaction through better organization; and (c) increased quality, productivity, and effectiveness. Kai is defined as to break apart or disassemble so that one can begin to understand. Zen is defined as to improve. This process focuses on improvements objectively by breaking down the processes in a clearly defined and understood manner so that wastes are identified, improvement ideas are created, and wastes are both identified and eliminated. The philosophy includes reducing cycle times and lead times in turn increasing productivity, reducing Work-In-Process (WIP), reducing defects, increasing capacity, increasing flexibility, and improving layouts through visual management techniques.

Operator cycle times need to be understood in order to reduce the nonproductive times. Operators should also be cross-functional so that they are able to perform different job functions and the workloads of each function are well balanced. The work performed needs to be not only value-added work, but also work that is in demand through customers. WIP should be eliminated to reduce inventory. Inventory should be seen simply as money waiting in process and should be reduced as much as possible. WIP can be reduced by reducing setup times, transporting smaller quantities of batch outputs, and line balancing. Bottlenecks should be removed by finding non-value-added tasks and removing the excess time spent by both machinery and humans. Sometimes, an eighth waste is added in and an abbreviation of DOWNTIME is associated with the acronym. It is defined as follows:

- Defect/Correction
- Overproduction
- Waiting
- Not utilizing employee talents
- Transportation/material movement
- Inventory
- Motion
- Excessive processing.

The primary technique for reducing waste and defects utilizing Lean Six Sigma techniques is demonstrated below in order of operations:

1. Define the problem.
   Root cause analysis by defining the problem.
2. Process mapping.
3. Data gathering – Gather data on any process with defects and issues. Utilize voice of the customer to find what data is needed.
4. Cause/effect analysis (seeking root cause) utilizing 5m's.
5. Verifying root cause with data-driven results – Ask why five times to ensure the proper root cause is found. Do not band-aid problems; instead, eliminate the cause that creates them.
6. Solutions and continuous improvement plans.
7. Test implementation plan by piloting. Pilot plans can include actual trials or mock trials with details information.
8. Implement continuous improvement ideas.
9. Control/monitoring plan.
10. Documentation of lessons learned.

Now that the process is laid out in terms of making proper improvements, the sustainability portion must be realized. The control plan during this phase is crucial. The control plan must not just be documented, but should be a living document that is followed structurally. The accountability portion for this phase is a key portion in having lasting results. The more specific the plan is, the better off the implementation of it will be. The plans must also be attainable; otherwise, the plan will fail.

Lean can also involve some statistical tools. The tools demonstrate the efficiencies and labor balancing. The main statistical tools are as follows:

First pass yield (FPY) indicates the number of good outputs from a first pass at a process or step. The formula is as follows:

FPY = (# accepted)/(# processed)

The formula for the FPY ratio is % FPY = [(# accepted)/(# processed)] × 100. This number does not include re-worked product that was previously rejected.

Rolled throughput yield (RTY) covers an entire process. If a process involves three activities with FPYs of 0.90, 0.94, and 0.97, the RTY would be 0.90×0.94×0.97=0.82. The %RTY=0.82×100=82%.

Value-added time (VAT) is % VAT=(sum of activity times)/(lead time)×100. When the sum of activity times equals lead time, the VAT is 100%. For most processes, % VAT=5 to 25%. If the sum of activity times equals the lead time, the time value is not acceptable and activity times should be reduced.

Takt time is a kaizen tool used in the order taking phase. *Takt* is a German word for pace. Takt time is defined as time per unit. This is the operational measurement to keep production on track. To calculate Takt time, the formula is time available/production required. Thus, if a required production is 100 units per day and 240 minutes are available, the Takt time=240/100 or 2.4 minutes to keep the process on track. Individual cycle times should be balanced to the pace of Takt time. To determine the number of employees required, the formula is (labor time/unit)/Takt time. Takt in this case is time per unit. Takt requires visual controls and helps reduce accidents and injuries in the workplace. Monitoring inventory and production WIP will reduce waste or muda. Muda is a Japanese term for waste where waste is defined as any activity that consumes some type of resource but is non-value-added for the customer. The customer is not willing to pay for this resource because it is not benefiting them. Types of muda include scrap, re-work, defects, mistakes, and excess transport, handling, or movement.

The Lean house is a common methodology for understanding Lean and waste reduction.

Mistake proofing is a subject of its own when brought into the Lean Six Sigma methodology. This term is often called Poka Yoke also known as another initiative to improve production systems. The methodology eliminates product defects before they occur by simply installing processes to prevent the mistakes from happening in the first place. These mistakes that happen are due to human nature and can normally not be eliminated by simple training or standard operating standard operating procedures (SOPs). These steps to eliminate the defect will prevent the next step in the process from occurring if a defect is found. Normally, there is some type of alert that will show there is a mistake and will fail the process from going forward. An example of a Poka Yoke would be a simple check weigher that would kick off a package of food if it were not the correct weight.

Poka Yokes often also encompass a concept called Zero Quality Control (ZQC). This does not mean a reduction in defects, but instead complete elimination of defects, also known as zero defects. ZQC was another concept led by the Japanese that leads to low-inventory production. The reason the inventory is so low is due to not needing excess inventory due to having to replace less defective parts as often. ZQC also focuses on quality control and data versus blaming humans on mistakes. The methodology was developed by Shigeo Shingo who knew it was human nature to make common mistakes and did not feel people should be reprimanded for them. Shingo said, "Punishment makes people feel bad, it does not eliminate defects."

This concept is important because it focuses on the customers and realizes that defects are costly, therefore eliminating defects saves money. Many companies "re-work" product to save money, but do not realize to eliminate the problem in the first place. This process will eliminate re-work by eliminating any defects from happening in the first place.

This is a traditional cycle where processes and conditions are planned out, the planned actions are performed in the Do phase, and finally quality control checks are performed in the Check phase. This method catches mistakes and also provides feedback during the Check phase. The checks in this place also account for 100% inspection therefore all parts or processes are looked upon indicating no defects.

There are three main types of checks or inspections that are popular:

- Judgment Inspections
- Informative Inspections
- Source Inspections.

Judgment Inspections are those that are done normally by humans based on what their expectations are. They find the defect after the defect has already occurred. Informative Inspections are based on statistical quality control (SQC), checks on each product, and self-checks. These inspections help reduce defects, but not eliminate them completely. Finally, the source inspections are the inspections that reduce the defects completely. Source inspections discover the mistakes before processing and then provide feedback and corrective actions so that the process has zero defects. The source inspections require 100% inspection. The feedback loop is also very quick so that there is minimal waiting time.

How to Use Poka Yokes:
Poka Yokes use two approaches:

- Control systems
- Warning systems.

Control systems stop the equipment when a defect or un-expected event occurs. This prevents the next step in the process to occur so that the complete process is not performed. Warning systems signal operators to stop the process or address the issue at the time. Obviously, the first of the two prevents all defects and has a more ZQC methodology because an operator could be distracted or not have time to address the problem. Control systems often also use lights or sounds to bring attention to the problem that way the feedback loop again is very minimal.

The methods for using Poka Yoke systems are as follows:

- Contact methods
- Fixed – value methods
- Motion – step methods.

Contact methods are simple methods that detect whether or not products are making physical or energy contact with a sensing device. Some of these are commonly known as limit switches where the switches are connected to cylinders and pressed in when the product is in place. If a screw is left out, the product does not release to the next process. Other examples of contact methods are guide pins.

Fixed-value methods are normally associated with a particular number of parts to be attached to a produce or a fixed number of repeated operations occurring at a particular process. Fixed-value methods utilized devices as counting mechanisms. The fixed-value methods may also use limit switches or different types of measurement techniques.

Finally, the motion-step method senses if a motion or step in the process has occurred in a particular amount of time. It also detects sequencing by utilizing tools such as photoelectric switches, timers, or barcode readers.

The conclusion of Poka Yokes is to use the methodology as mistake proofing for ZQC to eliminate all defects, not just some. The types of Poka Yokes do not have to be complex or expensive, just well thought out to prevent human mistakes or accidents.

The Poka Yoke discussion stems into the correct location discussion. This technique places design and production operations in the correct order to satisfy customer demand. The concept is to increase throughput of machines ensuring the production is performed at the proper time and place. Centralization of areas helps final assemblers, but the most common practice to be effective is to unearth an effective flow. U-shaped flows normally prevent bottlenecks. Value stream mapping is a key component during this time in order to establish all steps occurring are adding value. A reminder for value-added activities: any activity that the customer is willing to pay for.

Another note to remember is to not only have a smart and efficient technique, but also only produce goods that the customer is demanding to eliminate excess inventory.

This technique is called the pull technique. Pull is the practice of not producing any goods upstream if the downstream customer does not need it. The reason this is a difficult technique is because once an efficient method is found to produce a good, the mass production begins. The operations forget if the goods are actually needed or not and begin thinking only of through-put. Even though co-manufacturers seem like a bad idea for many employers, they sometimes come in handy when needed a small amount of a versatile product.

Push systems on the other hand are not effective due to predictions of customer demands.

Lean systems show the pull system utilizing machinery for 90% of requirements and limits downtime to 10% for changeovers and maintenance. This does not mean preventative maintenance should not be performed, but only that the maintenance time is reduced to 10%. Kanbans are a key factor during this Lean system in order to use a visual indicator that another part or process is required. This also prevents excess parts from being made or excess processes being performed.

Heijunka is the leveling of production and scheduling based on volume and product mix. Instead of building products according to the flow of customer orders, this technique levels the total volume of orders over a specific time so that uniform batches of different product mixes are made daily. The result is a predictable matrix of product types and volumes. For heijunka to succeed, changeovers must be managed easily. Changeovers must be as minimally invasive as possible to prevent time wasted because of product mix. Another key to heijunka is making sure that products are needed by customers. A product should not be included in a mix simply to produce inventory if it is not demanded by customers. Long changeovers should be investigated to determine the reason and devise a method to shorten them. The Lean action plan is simply drawn out in five steps:

1. Getting started – plan out the appropriate steps. This will take one to six months.
2. Create the new organization and restructure. This will take six to twenty-four months.
3. Implement Lean techniques and systems and continually improve. This will take two to four years.
4. Complete the transformation. This will take up to five years.
5. Do the entire process again to have another continuous improvement project and sustain the results.

ANALYTICS OF LEARNING CURVES FOR PROJECT INTEGRATION

No lasting benefits can happen without an opportunity to learn and improve. Improvements happen in project management through repeated actions of performing tasks. In its simplest form, learning curves measure how labor hours decline as proficiency increases due to repeated performance. With this, project managers can forecast how labor hours and cost will increase with higher levels of production. While the total cost will increase with an increase in output, the cost per unit will decrease due to the effect of learning.

The technique of learning curve analysis provides a good basis for the analytics required for project control. Degradation of performance occurs naturally either due to internal processes or externally imposed events, such as extended production breaks. For example, the COVID19 pandemic and the ensuing lockdown is an imposed scenario that adversely affects future performance and, ultimately, the ability of a project to produce the expected service, product, or result. The methodologies of learning curves and learn-forget curves can provide additional insights into a project environment. Although empirical data is not yet available for modeling post-COVID learning curves, a question that is relevant now is how COVID-19 affects learning. Further retrospective research along this line will be needed as the pandemic unfolds further.

Learning curves are important for resource allocation decisions (Badiru, 1996). Learning curves present the relationship between cost (or time) and level of activity on the basis of the effect of learning. An early study disclosed the "80% learning" effect, which indicates that a given operation is subject to a 20% productivity improvement each time the activity level or production volume doubles. A learning curve can serve as a predictive tool for obtaining time estimates for tasks in a project environment. Typical learning rates that have been encountered in practice range from 70% to 95%. A learning curve is also referred to as a *progress function*, a *cost–quantity relationship*, a *cost curve*, a *product acceleration curve*, an *improvement curve*, a *performance curve*, an *experience curve*, and an *efficiency curve*.

Several alternate models of learning curves have been presented in the literature. Some of the most notable models are the *log-linear model*, the *S-curve*, the *Stanford-B model*, *DeJong's learning formula*, *Levy's adaptation function*, *Glover's learning formula*, *Pegels' exponential function*, *Knecht's upturn model*, and *Yelle's product model*. The univariate learning curve expresses a dependent variable (e.g., production cost) in terms of some independent variable (e.g., cumulative production). The log-linear model is by far the most popular and most used of all the learning curve models. The log-linear model states that the improvement in productivity is constant (i.e., it has a constant slope) as output increases. There are two basic forms of the log-linear model: the average cost model and the unit cost model. The average cost model is used more than the unit cost model. It specifies the relationship between the cumulative average cost per unit and cumulative production. The relationship indicates that cumulative cost per unit will decrease by a constant percentage as the cumulative production volume doubles. The model is expressed as

$$A_x = C_1 x^b$$

where

A_x is the cumulative average cost of producing x units.
C_1 is the cost of the first unit.
x is the cumulative production count.
b is the learning curve exponent (i.e., constant slope of on log–log paper).
    The relationship between the learning curve exponent, b, and the learning rate percentage, p, is given by

$$b = \frac{\log p}{\log 2}$$

$$p = 2^b$$

The derivation of the previous relationship can be seen by considering two production levels where one level is double the other, as shown next.
    Let level I=x_1 and level II=$x_2=2x_1$. Then,

$$A_{x_1} = C_1 \left(x_1\right)^b$$

$$A_{x_2} = C_1 (2x_1)^b$$

The percent productivity gain is then computed as

$$p = \frac{C_1 (2x_1)^b}{C_1 (x_1)^b} = 2^b$$

When linear graph paper is used, the log-linear learning curve is a hyperbola. On log–log paper, the model is represented by the following straight line equation:

$$\log A_x = \log C_1 + b \log x$$

where b is the constant slope of the line. It is from this straight line that the name *log-linear* was derived. As an example, assume that 50 units of an item are produced at a cumulative average cost of $20 per unit. Suppose we want to compute the learning percentage when 100 units are produced at a cumulative average cost of $15 per unit. The learning curve analysis would proceed as follows:

Initial production level = 50 units; average cost = $20

Double production level = 100 units; cumulative average cost = $15

Using the log relationship, we obtain the following equations:

$$\log 20 = \log C_1 + b \log 50$$

$$\log 15 = \log C_1 + b \log 100$$

Solving the equations simultaneously yields

$$b = \frac{\log 20 - \log 15}{\log 50 - \log 100} = -0.415$$

Thus,

$$p = (2)^{-0.415} = 0.75$$

That is a 75% learning rate. In general, the learning curve exponent, b, may be calculated directly from actual data or computed analytically. That is,

$$b = \frac{\log A_{x_1} - \log A_{x_2}}{\log x_1 - \log x_2} \left(\text{for the case where two learning curve data points are available} \right)$$

$$b = \frac{\ln(p)}{\ln(2)} \left(\text{for the case where } p, \text{ the percentage learning, is known} \right)$$

where
x_1 is the first production level.
x_2 is the second production level.
A_{x_1} is the cumulative average cost per unit at the first production level.
A_{x_2} is the cumulative average cost per unit at the second production level.
p is the learning rate percentage.

Using the basic cumulative average cost function, the total cost of producing x units is computed as

$$TC_x = (x)A_x = (x)C_1x^b = C_1x^{(b+1)}$$

The unit cost of producing the xth unit is given by

$$U_x = C_1x^{(b+1)} - C_1(x-1)^{(b+1)}$$

$$= C_1x\left[x^{(b+1)} - (x-1)^{(b+1)}\right]$$

The marginal cost of producing the xth unit is given by

$$MC_x = \frac{d[TC_x]}{dx} = (b+1)C_1x^b$$

As another analytical example, suppose in a production run of a certain product, it is observed that the cumulative hours required to produce 100 units is 100,000 hours with a learning curve effect of 85%. For project planning purposes, an analyst needs to calculate the number of hours spent in producing the 50th unit. Following the notation used previously, we have the following information:

$p = 0.85$

$X = 100$ units

$A_x = 100,000$ hours/100 units $= 1,000$ hours/unit

Now,

$0.85 = 2^b$

Therefore, $b = -0.2345$
    Also,
    $100,000 = C_1(100)^b$
    Therefore, $C_1 = 2,944.42$ hours. Thus,

$C_{50} = C_1(50)^b = 1,176.50$ hours.

That is, the cumulative average hours for 50 units is 1,176.50 hours. Therefore, cumulative total hours for 50 units $= 58,824.91$ hours. Similarly,

$C_{49} = C_1(49)^b = 1,182.09$ hours.

That is, the cumulative average hours for 49 units is 1,182.09 hours. Therefore, cumulative total hours for 49 units = 57,922.17 hours. Consequently, the number of hours for the 50th unit is given by

58,824.91 hours – 57,922.17 hours = 902.74 hours.

The unit cost model is expressed in terms of the specific cost of producing the xth unit. The unit cost formula specifies that the individual cost per unit will decrease by a constant percentage as cumulative production doubles. The formulation of the unit cost model is presented as follows. Define the average cost as A_x.

$$A_x = C_1 x^b$$

The total cost is defined as

$$TC_x = (x)A_x = (x)C_1 x^b = C_1 x^{(b+1)}$$

and the marginal cost is given by

$$MC_x = \frac{d[TC_x]}{dx} = (b+1)C_1 x^b$$

This is the cost of one specific unit. Therefore, define the marginal unit cost model as

$$U_x = (1+b)C_1 x^b$$

U_x is the cost of producing the xth unit. To derive the relationship between A_x and U_x,

$$U_x = (1+b)C_1 x^b$$

$$\frac{U_x}{(1+b)} = C_1 x^b = A_x$$

$$A_x = \frac{U_x}{(1+b)}$$

$$U_x = (1+b)A_x$$

To derive an expression for finding the cost of the first unit, C_1, we will proceed as follows. Since $A_x = C_1 x^b$, we have the following expressions:

$$C_1 x^b = \frac{U_x}{(1+b)}$$

$$C_1 = \frac{U_x x^{-b}}{(1+b)}$$

For the case of continuous product volume (e.g., chemical processes), we have the following corresponding expressions:

$$TC_x = \int_0^x U(z)\,dz\, C_1 \int_0^x z^b\,dz = \frac{C_1 x^{(b+1)}}{b+1}$$

$$Y_x = \left(\frac{1}{x}\right)\frac{C_1 x^{(b+1)}}{b+1}$$

$$MC_x = \frac{d[TC_x]}{dx} = \frac{d\left[C_1 x^{(b+1)}/b+1\right]}{dx} = C_1 x^b$$

As another example, suppose an observation of the cumulative hours required to produce a unit of an item was recorded at irregular intervals during a production cycle. The recorded observations are used to develop a learning curve model for prediction purposes. The project analyst would like to perform the following computational analyses:

1. Calculate the learning curve percentage when cumulative production doubles from 10 to 20 units.
2. Calculate the learning curve percentage when cumulative production doubles from 20 to 40 units.
3. Calculate the learning curve percentage when cumulative production doubles from 40 to 80 units.
4. Calculate the learning curve percentage when cumulative production doubles from 80 to 160 units.
5. Compute the average learning curve percentage for the given operation.
6. Estimate a standard time for performing the given operation if the steady production level per cycle is 200 units.

A regression model fitted to the data is expressed mathematically as

$$A_x = 634.22 x^{-0.8206}$$

with an R^2 value of 98.6%. Thus, we have a highly significant model fit. The fitted model can be used for estimation and planning purposes. Time requirements for the operation at different production levels can be estimated from the model. From the model, we have an estimated cost of the first unit as

$C_1 = \$634.22$

$b = -0.8206$

$p = 2^{(-0.8206)} = 56.62\%$ learning rate

By using linear interpolation for the recorded data, we can estimate the percentage improvement from one production level to another. For example, when production doubles from 10 to 20 units, we obtain an estimated cumulative average hours of 60.6 hours by interpolating between 71.2 and 50 hours. Similarly, cumulative average hours of 13.04 hours was obtained for the production level of 80 units, and cumulative average hours of 7.75 hours was obtained for the production level of 160 units. Now, these average hours are used to compute the percent improvement over the various production levels. For example, the percentage improvement when production doubles from 10 to 20 units is obtained as $p = 60.6/92.5 = 65.5\%$. The calculated percent improvement levels are computed. The average percent is found to be 53.76%. This compares favorably with the 56.62% suggested by the fitted regression model. Using the fitted model, the estimated cumulative average hours per unit when 200 units are produced is estimated as

$A_x = 634.22(200)^{-0.8206}$

$= 8.20\,\text{hr}.$

Caution should be exercised in using the fitted learning curve for extrapolation beyond the range of the data used to fit the model.

MULTIVARIATE LEARNING CURVES

Extensions of the single factor learning curve are important for realistic analysis of productivity gain. In project operations, several factors can intermingle to affect performance. Heuristic decision-making, in particular, requires careful consideration of qualitative factors. There are numerous factors that can influence how fast, how far, and how well a worker learns within a given time span. Multivariate models are useful for performance analysis in project planning and control. One form of the multivariate learning curve is defined as

$$A_x = K \prod_{i=1}^{n} c_i x_i^{b_i}$$

where

A_x is the cumulative average cost per unit based on production level of x units.

K is the cost of first unit of the product.

x_i is the specific value of the ith factor.

n is the number of factors in the model.

c_i is the coefficient for the ith factor.

b_i is the learning exponent for the ith factor.

A bivariate form of the model is presented as follows:

$$C = \beta_0 x_1^{\beta_1} x_2^{\beta_2}$$

where

C is a measure of cost.

x_1 and x_2 are independent variables.

Using World War II data, researchers have experimented with alternate mathematical functions to estimate direct labor per pound of airframe needed to manufacture the Nth airframe in a cumulative production of N planes. The functions presented hereafter describe the relationships between direct labor per pound of airframe (m), cumulative production (N), time (T), and rate of production per month (DN):

1. $\log m = a_2 + b_2 T$
2. $\log m = a_3 + b_3 T + b_4 DN$
3. $\log m = a_4 + b_5 (\log T) + b_6 (\log DN)$
4. $\log m = a_5 + b_7 T + b_8 (\log DN)$
5. $\log m = a_6 + b_9 T + b_{10} (\log N)$
6. $\log m = a_7 + b_{11} (\log N) + b_{12} (\log DN)$.

The multiplicative power function, often referred to as the Cobb–Douglas function, has also been investigated as a model for learning curves. The model is of the general form:

$$C = b_0 x_1^{b_1} x_2^{b_2} \dots x_n^{b_n} \varepsilon$$

where

 C is the estimated cost.

 b_0 is the model coefficient.

 x_i is the ith independent variable ($i = 1, 2,..., n$).

 b_n is the exponent of the ith variable ($i = 1, 2,..., n$).

 ε is the error term.

 For parametric cost analysis, we can use an additive model of the form:

$$C = c_1 x_1^{b_1} + c_2 x_2^{b_2} + \cdots + c_n x_n^{b_n} + \varepsilon$$

where c_i ($i = 1, 2,..., n$) is the coefficient of the ith independent variable. The model was reported to have been fitted successfully for missile tooling and equipment cost. A variation of the power model was used to study weapon system production. The bivariate model has also been used for the assessment of the costs and benefits of a single-source versus multiple-source production decision with variations in quantity and production rate in major Department of Defense (DOD) programs. The multiplicative power model seems to be effective for expressing program costs in terms of cumulative quantity and production rate in order to evaluate contractor behavior.

 There is also a nonlinear cost-volume-profit model for learning curve analysis. The nonlinearity in the model is effected by incorporating a nonlinear cost function that expresses the effects of employee learning. The profit equation for the initial period of production for a product subject to the usual learning function is expressed as

$$P = px - c\left(ax^{b+1}\right) - f$$

where

 P is the profit.

 p is the price per unit.

 x is the cumulative production.

 c is the labor cost per unit time.

 f is the fixed cost per period.

 b is the index of learning.

 The profit function for the initial period of production with n production processes operating simultaneously is given as follows:

$$P = px - nca\left(\frac{x}{n}\right)^{b+1} - f$$

where x is the number of units produced by n labor teams consisting of one or more employees each. Each team is assumed to produce x/n units. This model indicates that when additional production teams are included, more units are produced over a given time period. However, the average time for a given number of units increases because more employees are producing, while they are still learning. That is, more employees with low (but improving) productivity are engaged in production at the same time. The preceding model is extended to the case where employees with different skill levels produce different learning parameters between production runs. This is modeled as follows:

$$P = p \sum_{i=1}^{n} x_i - c \sum_{i=1}^{n} a_i x_i^{b_i+1} - f$$

where

a_i and b_i denote the parameters applicable to the average skill level of the ith production run.

x_i represents the output of the ith run in a given time period.

This model could be useful for multiple projects that require concurrent execution. Multivariate learning curve models are available for incorporating cumulative production, production rate, and program cost. The approach involves a production function that relates output rate to a set of inputs with variable utilization rates. A bivariate model is used here to illustrate the nature and modeling approach for general multivariate models. An experiment conducted by the author models a learning curve containing two independent variables: *cumulative production* (x_1) and *cumulative training time* (x_2). The following model was chosen for illustration purposes:

$$A_{x_1 x_2} = K c_1 x_1^{b_1} c_2 x_2^{b_2}$$

where

A_x is the cumulative average cost per unit for a given set X, of factor values.
K is the intrinsic constant.
x_1 is the specific value of first factor.
x_2 is the specific value of second factor.
c_i is the coefficient for the ith factor.
b_i is the learning exponent for the ith factor.

Two data replicates are used for each of the ten combinations of cost and time values. Observations are recorded for the number of units representing double production volumes. The model is transformed to the natural logarithmic form:

$$\ln A_x = \left[\ln K + \ln(c_1 c_2) \right] + b_1 \ln x_1 + b_2 \ln x_2$$
$$= \ln a + b_1 \ln x_1 + b_2 \ln x_2$$

where a represents the combined constant in the model such that $a = (K)(c_1)(c_2)$. A regression approach yielded the fitted model:

$$\ln A_x = 5.70 - 0.21(\ln x_1) - 0.13(\ln x_2)$$

$$A_x = 298.88 x_1^{-0.21} x_2^{-0.13}$$

with an R^2 value of 96.7%. The variables in the model are explained as follows:
$\ln(a) = 5.70$ (i.e., $a = 298.88$)
x_1 is the cumulative production units.
x_2 is the cumulative training time in hours.

As in the univariate case, the bivariate model indicates that the cumulative average cost decreases as cumulative production and training time increase. For a production level of 1750 units and a cumulative training time of 600 hours, the fitted model indicates an estimated cumulative average cost per unit as

$$A_{(1750,600)} = (298.88)(1750^{-0.21})(600^{-0.13}) = 27.12$$

Similarly, a production level of 3,500 units and a training time of 950 hours yield the following cumulative average cost per unit:

$$A_{(3500,950)} = (298.88)(3500^{-0.21})(950^{-0.13}) = 22.08$$

To use the fitted model, consider the following problem. The standards department of a manufacturing plant has set a target average cost per unit of $12.75 to be achieved after 1000 hours of training. We want to find the cumulative units that must be produced in order to achieve the target cost. From the fitted model, the following expression is obtained:

$$\$12.75 = (298.88)\left(X^{-0.21}\right)(1000^{-0.13})$$

$$X = 46,409.25$$

On the basis of the large number of cumulative production units required to achieve the expected standard cost, the standards department may want to review the cost standard. The standard of $12.75 may not be achievable if there is a limited market demand (i.e., demand is much less than 46,409 units) for the particular product being considered. The relatively flat surface of the learning curve model as units and training time increase implies that more units will need to be produced in order to achieve any additional significant cost improvements. Thus, even though an average cost of $22.08 can be obtained at a cumulative production level of 3,500 units, it takes several thousand additional units to bring the average cost down to $12.75 per unit.

ANALYTICS OF INTERRUPTION OF LEARNING

Interruption of the learning process can adversely affect expected performance. An example of how to address this is the *manufacturing interruption ratio*, which considers the learning decay that occurs when a learning process is interrupted. One possible expression for the ratio is

$$Z = (C_1 - A_x)\frac{(t-1)}{11}$$

where
   Z is the per product loss of learning costs due to manufacturing interruption.
   $t = 1, 2, ..., 11$ (months of interruption from 1 to 12 months).
   C_1 is the cost of the first unit of the product.
   A_x is the cost of the last unit produced before a production interruption.
   The unit cost of the first unit produced after production begins again is given by

$$A_{(x+1)} = A_x + Z$$

$$= A_x + (C_1 - A_x)\frac{(t-1)}{11}$$

Interruption to the learning process can be modeled by incorporating forgetting functions into regular learning curves. In any practical situation, an allowance must be made for the potential impacts that forgetting may have on performance. Potential applications of the combined learning and forgetting models include design of training programs, manufacturing economic analysis, manpower scheduling, production planning, labor estimation, budgeting, and resource allocation.

INTEGRATING A CHECKLIST INTO WORK PROCESS

An account in the US national archives indicates that on October 30, 1935, a Boeing plane known as the "flying fortress" crashed during a military demonstration at Wright Field, a flying forerunner of the present-day

Wright-Patterson Air Force Base in Ohio. The shocking accident baffled the aviation industry of that era. It led to questions about the future of flight, in general, and the Boeing program, in particular. In response, the **checklist** was introduced by Boeing, as a permanent and mandatory tool, to be used by all pilots in the Boeing fleet. Thus emerged the industry standard of checklists. Even today, a pilot cannot take off or land without going through a prescribed checklist. The same techniques of aviation checklists can be adopted for project management applications and, specifically, for systems integration.

As presented by Ludders and McMillan (2017), the history behind the checklist is intriguing and inspirational. Having its origin linked to the military makes it even more of interest to this author, who presently works at Wright-Patterson Air Force Base. Like many technical advances, the safety checklist has its origin in the military. The popular project management techniques of CPM and PERT originated and spread on the basis of military needs of the past (Badiru, 1996). In 1935, the US Army Air Corps started a final set of aircraft evaluations at Wright Field, Dayton, Ohio. On the line was a contract to supply the US Army with potentially up to 200 long-range bomber aircraft. There were three aircraft competing for this large and lucrative contract, one of which was the Boeing Model 299. It was believed that all the initial evaluations (consisting of about 40 hours of flight time) had gone in favor of Boeing favor. The final test flight was expected to be a mere formality. Boeing's entry had already earned itself the nickname of "the Flying Fortress" because it could carry considerably more bombs and fly faster and farther than the other two entries. Flying the Model 299 that day were two highly experienced Army pilots, Boeing's chief test pilot, along with a Boeing mechanic and a representative of the engine manufacturer. After takeoff, the Model 299 began to climb, but within a few seconds, the aircraft stalled and fell to the ground, bursting into flames upon impact. Although all onboard escaped or were rescued, both pilots later died of their injuries. Compared to a typical plane at the time, the Model 299 was a complex aircraft with additional controls and instruments that required attention. Finding no evidence of mechanical malfunction, the accident investigation team assigned to the crash concluded that "pilot error" was the cause. Evidently, the pilots had made a simple mistake with one of the new controls, leaving the elevator and rudder controls locked. A newspaper at the time went on to state that the Model 299 was just "too much plane for one man to fly." This could have been the end of the story, but for the huge potential advantage, the bomber would give the US army if it could be flown safely. So, although the main contract was for the Douglass DB-1, a dozen Model 299 planes were purchased for testing purposes. After some deliberation, the solution to the problem was simple, ingenious, but most of all effective: the pilots' checklist. It turned out the plane was not too much for one man, but merely too much for one man's memory; a simple checklist could ensure that none of the crucial steps during the key periods of flight were forgotten. Four checklists were initially developed: takeoff, flight, before landing, and after landing. All pilots were taught how to use the checklist as part of their normal flight training. The initial 12 Model 299s tested by the army went on to fly almost 2 million miles without serious incident and the army went on to order over 10,000. The army renamed the aircraft B-17, and it became an icon, a symbol of power for the US Air Force. The checklist idea was so successful that it enabled aviation and aeronautical engineering to become more and more complex. Checklists were developed for more and more parts of flight, for emergency situations as well as more routine situations. As an example, checklists were developed for almost every part of the Apollo missions and all astronauts were trained in how to use them and write them. Each of the Apollo 11 astronauts logged over 100 hours of time familiarizing themselves with and adapting these checklists. In fact, checklists were so integral to the success of the Apollo moon landings that astronaut Michael Collins coined them "The fourth crew member." Nowadays, every project needs a fourth angle of support, possibly through a systems-informed checklist.

In a general application, the major function of the checklist is to ensure that the team performs every and all required activities in the prescribed consistent sequence. A checklist can be effective for the following:

1. Aid the project team in recalling the elements contained in a work process.
2. Provide a standard foundation for verifying the sequence of activities.
3. Reduce the mental, physical, and psychological stress on individuals in the team.
4. Provide a consistent control of functions.
5. Provide a reliable framework for operational requirements.
6. Allow mutual supervision and cross-checking of activities among team members.
7. Facilitate keeping all team members in the loop of situational awareness.
8. Explicitly identify the duties of each participant in the project.
9. Pave the way for project communication, cooperation, and coordination.
10. Serve as a quality control tool for project execution.
11. Permits equitable distribution of work.
12. Support the ideals of Lean operations and Six Sigma outputs.

CONCLUSIONS AND RECOMMENDATIONS

Chapter 5, as the concluding chapter for this book, brings everything together under the platform of integration. Project design, evaluation, justification, and integration ensure that a project's goals and objectives remain in focus throughout the project effort. This chapter combines the methodology of SMART operations with the expectations of the DEJI Systems Model to boost the relevance of the systems approach to project management. The section on checklists is particularly of high utility as a guide for project execution. As with any tool, technique, methodology, or model, the proof of the approach will be in the real-world usage of the approach, where actual utility will be customized to the prevailing project environment and any unique circumstances. There are different strokes for different projects. In each case, the project team must demonstrate attention to detail, such as that provided by the checklist approach.

Since time is of the essence in any project management effort, a summary recommendation in this book is that if two options, one with a short duration and one with a long duration, would both yield the same end result, everything else being equal, the shorter-duration option should be selected. This might appear to be counter-intuitive, but consider a case where lecturing for 30 minutes or lecturing for two hours would, equally, produce the same learning outcome. Obviously, the shorter 30-minute lecture should be selected so that the 90 minutes saved could be put to use for other productive project pursuits. We rest our case on this note.

REFERENCES

Badiru, A. & Kovach, T. (2012). *Statistical techniques for project control*. New York: Taylor & Francis Group.

Badiru, A. B. (1996). *Project management in manufacturing and high technology operations* (2nd ed.). New York: John Wiley & Sons.

Ludders, J. W. & McMillan, M. (2017). *Errors in veterinary anesthesia* (1st ed.). Hoboken, NJ: John Wiley & Sons, Inc.

Index

Printed in the United States
by Baker & Taylor Publisher Services